LIVES OF WEEDS

LIVES OF WEEDS

Opportunism, Resistance, Folly

JOHN CARDINA

COMSTOCK PUBLISHING ASSOCIATES
AN IMPRINT OF
CORNELL UNIVERSITY PRESS
ITHACA AND LONDON

First published 2021 by Cornell University Press

Printed in the United States of America

Library of Congress Cataloging-in-Publication Data

Names: Cardina, John, 1953– author.
Title: The lives of weeds : opportunism, resistance, folly / John Cardina.
Description: Ithaca [New York] : Comstock Publishing Associates, an imprint of Cornell University Press, 2021. | Includes bibliographical references and index.
Identifiers: LCCN 2020052565 (print) | LCCN 2020052566 (ebook) | ISBN 9781501758980 (paperback) | ISBN 9781501759000 (ebook) | ISBN 9781501758997 (pdf)
Subjects: LCSH: Weeds. | Plants and civilization. | Weeds—History.
Classification: LCC SB611 .C37 2021 (print) | LCC SB611 (ebook) | DDC 632/.5—dc23
LC record available at https://lccn.loc.gov/2020052565
LC ebook record available at https://lccn.loc.gov/2020052566

To James and Angelina, and all who have followed

CONTENTS

Preface   ix

Acknowledgments   xi

Introduction: Clearing a Path   1

1. Dandelion   17

2. Florida Beggarweed   45

3. Velvetleaf   73

4. Nutsedge   103

5. Marestail   131

6. Pigweed   157

7. Ragweed   185

8. Foxtail   211

Epilogue: What's 'Round the Bend   234

Notes   241

References   251

Index   275

PREFACE

I have one of the best jobs in agriculture: I study weeds. I teach people about weeds. I grow weeds in greenhouses as one might grow orchids. I visit them in fields and stand beside them in their senescence.

For the last thirty years I've been conducting agricultural research and education at the Ohio Agricultural Research and Development Center (OARDC), in Wooster, Ohio, a campus of the Ohio State University. Before that I worked with the USDA in Tifton, Georgia. Along the way, I've had the freedom to explore research questions on a topic that touches everyone.

Over time, my observations of weeds—and the people who are concerned, frustrated, or offended by weeds—changed the way I think about our relationship with them. I came to see weeds as an outcome of our interactions with the natural world. I found weeds entangled in beliefs, attitudes, and behaviors toward nature, gardening, food, and more. This book is my attempt to explore the long and ongoing relationship with weedy plants that have complicated our lives, and whose connection to our own history has long been overlooked.

The book is organized around eight individual (or group of related) weed species. I chose these particular weeds because I know them through my research and travel. They represent different ways human and plant behavior have led to weediness. The chapters are arranged in a sequence that roughly represents a trend in the scale and complexity of human-weed interactions. The order also corresponds to an evolving understating of my own complicity in those interactions.

Information in the chapters is based on published research as well as my own observations as a researcher, gardener, observer of nature, and reader of history. Accounts of my personal involvement represent my interpretation of events. I have done my best to recall conversations as I heard and understood them, based on notes and corroboration with those involved. Names of some people and places have been changed and a couple are composites. Quoted dialogue represents my best recollection of the meaning, if not the exact words. The order of events has sometimes been adjusted for a smoother narrative without altering the veracity of what transpired.

I have included some basic biology to explain how weedy plants function. Weed biology involves evolutionary biology, and genetics, and plant reproduction. And as herbicides are used to kill weeds, I explain in general how these things work—and how they often don't. I expect readers will find that the science surrounding the plants is no more difficult to comprehend than the responses and behaviors of the human participants in the stories.

ACKNOWLEDGMENTS

I am indebted to more people than I have space to mention or memory to recall. I would like to thank Kitty Liu at Cornell University Press for championing this project and encouraging my efforts to bring it about.

This book never would have gone beyond scrawled notes on scraps of paper without the love, support, and enthusiasm of my first, best, long-lost and finally found editor, Barbara Hoekje, whose close reading and kind suggestions on every draft of every chapter helped me see my way into the material and onto the page with new clarity.

I am incredibly fortunate to have entrusted my brother, Tim, with this text. His meticulous reading, editorial suggestions, and corrections on history and medical issues were always delightful. Infinite thanks to Caitlin and Mollie for loving, artful, daughterly advice. I thank many other helpful readers, especially Daniel Olivieri and James F. Sassaman, who took on the challenge of unfamiliar subject matter.

I spent many hours on country roads and in fields with John Holliger, poet and landscape photographer, whom I convinced to turn a lens toward weedy plants. His approach to art and life helped me see and understand the weedy world more clearly. I thank him for allowing me to include the images that appear in these pages.

I acknowledge the work of the wise people who digitized and made available thousands of primary sources of botanical literature which I have drawn on generously. I especially thank librarians at Ohio State University, including Connie Britton, Florian Diekmann, Laura Miller, and Gwen Short; Duncan McClusky at the Georgia Coastal Plain Experiment Station Library; and the helpful staff at the Philadelphia Academy of Natural Science as well as the New York Botanical Garden.

I deeply regret that I could not share this writing with Ted Webster and Ben Stinner, two colleagues whose friendship and perspectives I have tried to represent faithfully. Their reading and insights would have made this a better book.

Many academic colleagues have provided ideas and insights that led me to look at weeds in new ways. I especially thank Jim Metzger, chair of the Department of Horticulture and Crop Science at Ohio State University, for allowing me the freedom to work in places and ways that fit my schedule and approach to writing. Research and ideas on the interactions between weeds and people were inspired by many, including M. K. Antwi, Katrina Cornish, Adam Davis, Jack Dekker, Toni DiTommaso, Doug Doohan, M. K. Dzasimatu, Frank Forcella, Jonathan Fresnedo-Ramirez, Kent Harrison, Dan Herms, Casey Hoy, Zahid Hussain, Parwinder Grewal, George Kegode, Matt Kleinhenz, David Kline, Ramon Leon, Mark Loux, Ed McCoy, Richard Moore, Emilie Regnier, Debbie Stinner, and Charles Swann. Working with graduate students motivated this writing, and I hope did not hinder their careers, especially Lynn Sosnoskie, Stephanie Wedryk, Robert Gallagher, Jing Luo, Mark Thorne, and Brian Iaffaldano. I have been gifted with technical research support from the best research assistants anyone could have, in particular Catherine Herms, Denise Sparrow, Paul McMillen, and Steve Hansen.

I thank the many farmers, gardeners, and extension educators who have been my teachers and guides to practical issues and philosophical perspectives in ways that have been especially helpful, understanding, and patient. I apologize to anyone I have forgotten to list. For those mentioned in these stories, I ask forgiveness if my account of events is in any way faulty.

LIVES OF WEEDS

INTRODUCTION

Clearing a Path

So many troubles and anxieties of our time are entangled in lowly weeds. Big issues like food justice, environmental crises, and climate chaos have significant connections to plants of disrepute. Personal questions about your health, property values, and what to put on the end of your fork, are questions about unwelcome botanical companions. So a book about weeds is also about you and me and where we see ourselves in the natural world.

Our entanglement with weedy plants goes back thousands of years. Human activity has provided the conditions for the evolution of adaptive traits that made certain plants especially widespread, persistent, and troublesome. And humans have been especially effective agents of their global dispersal. The utterly human response to the presence of an obnoxious plant has frequently led to conditions that in some way favor the offending species or others like it.

The notion of paths is a shorthand way to describe human-plant interactions. These interactions involve plant evolutionary biology as influenced by human behaviors and technologies. Certain plant species followed particular paths to become successful weeds because humans, in turn, responded to them in ways that have favored their continued adaptation. Thus, ordinary, unobtrusive plants with inherited survival traits achieved ecological success and spread their weedy genes across diverse environments around the world. They couldn't have done so by themselves. Humans unintentionally provided selection pressures for the evolution and survival of traits for botanical rudeness. And the intrusion of unwanted plants has motivated people to change the way they think and behave toward the environment. As a result, the paths to weediness resemble coevolutionary relationships: weeds wouldn't be weeds without us, and we wouldn't be who we are without weeds.

I've chosen eight weeds to represent different paths made possible by various features of plant biology and human interactions. Most familiar is dandelion, which followed a path opened by the human imagination and its construction of social rules.[1] Dandelions exploited this opening with the aid of a unique seed dispersal mechanism and breeding system. Least familiar is Florida beggarweed, a southern species that followed a path made available by unintended dispersal facilitated by slavers, opportunists, and sticky seedpods. Velvetleaf became a troublesome plant in the footsteps of nation building along with entrepreneurial hopes that disregarded the power of seed ecology and biological plasticity. For nutsedge, the path has been one of poverty and neglect for a species that is both crop and weed with unique chemical properties. Marestail was an insignificant waif until industry genetically engineered a path to herbicide resistance that revealed a species with unanticipated potential for dispersal and growth. The path for pigweed was forged by a human tendency to make the same mistake over and over while getting the same result from a prolific species with unusual facility for genetic change. Ragweeds followed paths of environmental disturbance that led them out of riversides into crop fields and across the world with war and economic development where they exploited their ability to thrive on contaminated soils in a changing climate. Foxtail became a major weed when the expansion of American agriculture after World War II opened a path for a novel robust grass to cross the prairie; meanwhile, its grassy relatives continued as useful grain or forage crops or minor inconveniences, the balance among them controlled by users of the landscape.

These species, like all weeds, are plants of contradiction—despised and admired, useless and essential, targets for eradication and sources of useful genetics. They are products of the human ambivalence surrounding food, labor, and our relationship with the natural world. They spoil the efforts of every gardener, farmer, nature lover, homeowner, and anyone who ever planted a hopeful seed in the soil. In direct and unseen ways, they complicate the life of everyone who enjoys the flowers and fruits of the earth. In suburban North America, weeds might be regarded as a nuisance; in the Midwest they're a fixed cost in farming; for subsistence communities across the world, they're an existential challenge that commands the bulk of family labor. Yet some are beautiful, with practical uses, and many play critical roles in the function of ecosystems.

Adding to the confusion is the paradoxical concept of *cultivate*. We cultivate flowers or crops, meaning to plant and tend them for whatever purpose. To do so in a garden, we spend little time planting or sowing, and a lot of time weeding. Cultivate also means to remove weeds with a hoe or mechanical cultivator. We cultivate weeds to destroy them; we cultivate crops to nourish and foster them. The contrary meanings of "cultivate" have the same objective: destroying weeds so crops may thrive. We wouldn't do one without the other.

Weeds remain a source of bewilderment. You see an unfamiliar plant in a field or garden and the questions erupt: What is this plant? How did it get here? What does it mean that it is here, standing, creeping, or grassing in our way? And ultimately: What am I supposed to do about it? Removing one just makes room for another. Or maybe a dozen. Every new tool ever invented to control the usual suspects just emboldens the pricklier ones that cause more damage and are tougher to control. The weeds—the botanical bullies—always win. How can this be?

The answers to these questions are hidden along the paths that otherwise inconsequential plants have traveled to become weeds with the help of human accomplices. The paths differ from one weed to another. They all involve basic plant evolutionary biology tangled with human behavior. Plant biology involves breeding systems that generate genetic variation leading to inherited traits that act together with plasticity to favor plant survival. Human behavior involves ancient and modern practices to manipulate the environment so that crops may grow, flowers may bloom, and desirable plants may thrive. Inevitably, there is opportunism, as weeds, like their human companions, are opportunists that grab resources at the expense of others. Both exhibit resistance, which occurs in weeds through genetic changes, whereas humans resist changes that might make weeds more tolerable. Alas, only humans exhibit folly.

Plants to Know and See

Weeds are intriguing, in part because they're so hard to define. Books, essays, and philosophical tracts have been devoted to the task. Anybody who tends a garden recognizes them. Anybody who participates in the global food system is linked to their fruition and demise. Yet most definitions for weeds just don't satisfy.

People chuckle over Emerson's quip that weeds are plants "whose virtues have not yet been discovered."[2] Clever words. But culinary, medicinal, and practical virtues of most common and troublesome plants have generally been well known for centuries. Today's hybridized, polyploid, herbicide resistant, epigenetically altered species that cost billions of dollars yearly to control are probably not the plants that tickled the transcendentalist's musing or delighted the simplicity of his childhood.

Others are satisfied with the old chestnut that weeds are "plants out of place."[3] Yet these are plants that establish, survive, and reproduce in places that are uniquely theirs. They have no other place. Their place is often marked by soil disturbance, chemical farming, GMO crops, pesticides in groundwater, and soil erosion. In fact, troublesome plants arise where humans inadvertently select them and enhance their adaptation while trying to subdue them. For some weeds, their place lies in the eye of the beholder, people offended by them in spite of their inconsequentiality.

Many people have attempted to define weeds by ticking off a checklist of traits that make a plant intolerable. One commonly used list of "weedy traits" includes things like easy germination, persistent seeds, rapid growth, self-pollination, prolific seeding, adaptations for dispersal, and others.[4] Nobody knows how many of the traits are needed to make a plant a weed. Or if a plant can be a weed without any of the traits. Or have all the traits yet not be a weed. Veritable weeds resist such categorization. They're botanical nonconformists. They look and behave in different ways. One person's weed is another's wildflower, food, or medicine. For my part, your prickly thicket of frustration looks a lot like job security.

I haven't found a definition more useful than Justice Stewart's "I know it when I see it." Ambiguous, yes; but it's honest and functional. A plant is weed in relation to human values. That means economic interests, perceptions of beauty, and social norms. On the other hand, when a grower has invested thousands of dollars to sow and nurture a high-value crop and wakes to find the field infested with a robust mutant vegetal freak that has ensnared the crop and cannot be controlled, no philosophical ruminations are needed. The offending species is likely to be defined in clear, colorful, consonant-rich language that defies social norms pertaining to religion, sex, or bodily functions.

Maybe the idea of a "weed" is hard to define because "weed" is just that: an idea. In other words, human and plant behaviors give rise to plants regarded as loathsome in the eyes and hearts of those who consider them so. In this circular sociobotanical logic, plant biology and human culture are tangled together; they can't be unraveled. Maybe "weed" is hard to define because they're so like us. We see ourselves in them. They colonize precious spaces and exploit resources to make more of themselves. They are inherently intrusive, pushy, competitive, and obnoxious. Some smell bad. Some are prickly. Some are ugly. Weeds can be like that too.

The historical and evolutionary entanglement of humans and weeds fills a peculiar niche in the range of associations between unrelated species. I don't think it's too great a stretch to describe this human-plant interaction in terms of coevolution. In coevolution, two species evolve together, each changing in response to changes in the other. I use the term cautiously, somewhat beyond its technical definition. Coevolution results in evolutionary changes that would not have occurred without cross-species interactions. Humans and weeds aid and abet each other in changes in behaviors, adaptations, physiology, and genetics, which would not otherwise have occurred. Moreover, human attitudes, technologies, and behaviors have changed in response to the greater fitness and wider distribution of weeds.

The Path to Domesticity

The very idea of weedy plants came about long before anybody thought they needed a definition. Astute ancient people foraging for nuts and berries must have noticed that useful plants grew better in spaces free of other wild plants. Desirable plants provided food, medicine, shelter, and materials, while others did not. Still other plants interfered with efforts to produce and harvest the desirable ones. And with that, the concept of "weed" arose in the human consciousness. No longer just a wild plant to avoid while gathering fruits and nuts, a "weed" became something new, suspect, mysterious.[5]

As soon as the concept of "weed" was born, the fantasy of "weed removal" could not have been far behind. Without their removal, newly domesticated crops could not garner enough light, water, and nutrients to yield a worthy harvest. The removal imperative touched off a giant leap in the effort of humans to exert control over nature and the invention of technology to do so.

Most historians have overlooked the place of weeds in the great Agricultural Revolution, one of the greatest steps (or mistakes) in human history. In school we were taught that around twelve thousand years ago, at several places across the world, ingenious humans—mostly women—cultured plants and domesticated crops. People gave up wandering and gathering for a settled agriculture where a few select plants and animals became the focus of diet and culture. This led to civil society with social structure and other hallmarks of civilization, like traffic jams and social media platforms.

Some writers have suggested that settled agriculture began when wild ancestors of plants like wheat and potatoes essentially took charge of their fate. These plants provided attractive and tasty complex carbohydrates in exchange for human efforts to clear fields, sow, and harvest. The ecological success of the plants was assured when people carried and reproduced the seeds—thus the genes—of these plants worldwide. From this perspective, humans did not domesticate crops so much as crop plants domesticated humans, by inducing people to settle down and invent settled agriculture.[6]

But there's more to this story. In fact, it was weeds that led to human settlement and civilization. Those first ingenious farmers scattered seeds and buried shoots but returned after a few months to find fields full of unruly thistles and grasses, not harvestable crops. The essential feature that led to settled domestic living was the need to hunker down and engage in constant weed removal. From this arose an ancient method of soil cultivation, or "hoe culture," based on a novel technology: a bent stick or sharp rock whose purpose was to get rid of weeds. These rudimentary tools scratched the earth and cleared away the botanical misfits so desirable plants might thrive in the great Revolution.

Thus, it was weeds—not crops—that compelled people to settle down and demanded their repeated attention. Weeds—not crops—domesticated humans. Unlike well-behaved crop plants, weeds germinated whenever they felt like it, helped themselves to resources intended for the crop plants, refused to hold on to their seeds for humans to harvest, and scattered themselves wherever opportunities arose to grab more goods for themselves. Without sacrificing any complex carbohydrates or nutrient-filled fruits, weeds relegated humans to constant battle.

The significance of settled agriculture for human history can hardly be overstated. In many respects, it made us who we are. It also entangled humans along paths that made weeds what they are. For the next dozen or so millennia, the history of what became known as "progress" has been a story of the human struggle against weeds. And this struggle has made the leafy scoundrels more spiteful. The struggle has selected the species and genotypes that tolerate continual human efforts to get rid of them. The struggle endures.

Revolutions Have Consequences

Weed removal, control, management—whatever you call it—became part of the great Revolution's impact on people and the environment. There's no getting around it: the intentional growing of food disturbs the environment. It happens whenever a tool scrapes across the earth to bury a seed, harvest a tuber, or extract a wayward plant. Every approach ever invented for planting, tending, or harvesting food crops to feed civilizations causes some disturbance of the natural environment, as much as we might wish otherwise. As soon as the soil is opened, berries are collected, or the remains of an animal are brought to a hearth, the natural world has changed.

Settled agriculture, with its need for weed removal, presented new challenges for early societies. Preagricultural people had leaders and followers, hunters, collectors, herbalists, and arrow makers. But crop farming called for new divisions. The invention of weeds, in particular, demanded a new type of human activity—bending over, pulling, and digging weeds out of the ground. Expanded crop and animal production led to increasingly organized civic structures, and conditions were in place for agricultural peoples to thrive. All except for one thing: nobody wanted to be the one who spent their days in the sun bending over, pulling, and digging weeds out of the ground.

The Agricultural Revolution focused immense human resourcefulness on the development of new crops and methods to make them grow. When faced with the dilemma posed by weedy plants, ancient people dipped into this same font of creativity: instead of simply killing those who had been captured or abandoned by war, put them to use in the field doing the task nobody else wanted to do—bending over, pulling, and digging weeds out of the ground. Thus, the drudgery of weed removal became the burden of enslaved human beings in nearly every civilization across the world.[7]

Historians have not connected weeds to this part of agricultural history. The point of studying history, as it was explained to me, is to figure out how we got into our current mess. It seems to me that weeds and their control shouldn't be overlooked. A good case could be made that the Anthropocene—the geologic epoch defined by significant human altering of Earth—began when the first hoe touched the soil to clear away the primeval undergrowth. Many environmental and social stresses we face as the Anthropocene unfolds can be traced to the consequences of unsolicited plants in settled agriculture. Ancient clearing, burning, and hoeing prefigured widespread erosive losses of soil and modern-day impacts on air and water quality from weed-killing herbicides. Early attitudes about weeds and people who hoe them persist in prejudices toward different social classes.

Weeds link us back to ancient times: modern control measures, mechanization, and herbicides are essentially replacements for the labor—forced or otherwise—in row-crop agriculture. Interactions between people and plants that began in the Agricultural Revolution echo in the natural histories of the species described here. Whether knowing these histories will help us escape our complicity in the creation of weeds or alter our destiny is another question. It all depends, I suppose, on choices we make based on how we see ourselves in relation to the natural world.

Selection in the Field

Most of us know weeds as passing strangers. They're in and out of our lives. We hope mostly out. We think of them in a certain way as though they've always been that way. Over a lifetime of gardening or farming, our attention is on the vegetables, flowers, or grains. New varieties are released every year, whereas the weedy plants look the same. Yet the ill-favored flora has been changing—evolving—for a very long time. In fact, weeds are among the best examples of the ongoing—and sometimes rapid—evolution of plants.

When agriculture was first invented, wild plants faced new circumstances with new types of stress brought on by farming. Only a few of the wild species could withstand the annual earth scratching, burning, land clearing, soil mounding, and repeated hoeing. Species that couldn't tolerate these disturbances were pushed aside. That left more space and resources for the plants that tolerated—or thrived—where the soil was disturbed. Simple agricultural practices favored some species over others. Scraping the ground with a hoe after a rain favored wild seeds that were at the right depth and had the ability to germinate after a rain. Scraping the ground at other times, or in different ways, favored other plants. The simple act of pulling plants out of the soil by hand eliminated species having weak roots firmly attached to smooth stems. They were tossed aside. That left plants with strong root systems, the ability to break off at the surface, or those with prickly spines too painful to grab. It didn't matter what approach those early farmers used to cultivate weeds; other species of plants survived to replace the ones that were easy to remove.

What remained in fields were vagabond plants with strong roots, seed dormancy, and dispersal mechanisms that helped them survive. The tougher species endured. Whenever different crops were grown, animals grazed the fields, or soil was manipulated in particular ways, different plants were favored and became adapted to survive as botanical intruders. Thus, the paths to weediness are littered with species that would, could, and some day might be weeds if human practices created conditions in their favor.

As agriculture has gotten more sophisticated, so have weeds. Far beyond hoes scraped across the ground, farmers have spent the last ten to twelve thousand years putting hateful herbs through the agricultural meat grinder with practices that include tillage, changing soil fertility, crop competition, irrigation, and harvesting. It's the only way to grow flowers, gardens, grains, orchards, lawns, or other organisms that allow people to enjoy the wonders of photosynthesis. As the pace of mechanical, chemical, and genetic manipulation has increased, the stresses, or *selection pressures*, have changed. Evolution of plants works by favoring genes that code for adaptive survival traits. In the setting of modern agriculture, human behaviors and technologies have imposed most of the selection pressures that determine which genotypes are favored.

I am proposing the term *agrestal* (pronounced "uh-GRES-tal") *selection* to describe the way weedy plants best adapted to agricultural field environments survive to produce more, stronger offspring. The selection process operates just like natural selection. But it isn't "natural" because the environment is a highly human-altered agricultural setting. And the selection pressures—like tillage or herbicides—are distinctly unnatural. Nor is it "artificial" selection, the process used by plant breeders to intentionally select and mate plants to produce new crop varieties. Darwin wrote a lot about weeds, but he never invented a word for this type of selection. Evolutionary biologists like deWet and Harlan also wrote extensively about weeds, but they referred to this process as simply evolution in the "man-made" habitat.[8]

"Agrestal" is a durable old word for plants that grow wild in tamed agricultural fields. Thus, agrestal selection highlights the human role in evolutionary paths that create new and changing weed populations. Agrestal selection drives the ongoing evolution of weedy plants— genotypes selected for complicated seed dormancy, unsynchronized seed germination, irregular flowering, casual seed dispersal, and other traits that allow them to succeed in modern agriculture. Through agrestal selection, farmers and gardeners have inadvertently helped to increase the ecological success of weeds by favoring genes that allow them to tolerate mowing, cultivating, burning, and other physical controls that we employ thinking we will thwart their success. And agrestal selection is how repeated use of the same herbicide leads to the evolution of novel biotypes that are resistant to herbicides.

Agrestal selection links people to weeds and evolution through agricultural systems on which societies depend. The more effective our control tactics, the greater the agrestal selection pressure for genotypes that tolerate those tactics. In other words, agrestal selection happens in response to the best intended efforts to manage weeds. And the result is that, inevitably, some plant species will tolerate those efforts better than others, so they survive and reproduce more successfully as outcasts. This pattern of plant evolution in response to human activity is fundamental to the paths to weediness. It is part of the natural histories of all the species described here. This basic concept is all the knowledge of genetics you'll need to understand the rest of the story.

Choosers and the Chosen

The math on weed evolution seems hard to figure. There are 400,000 species of flowering plants, give or take a few thousand. About 300 or so have been labeled weeds at some place or time. Of these, a couple dozen reign today as remarkably successful regional or worldwide pests of economic and practical significance. Yet many traits that are useful for plant survival are not so rare among would-be weeds lurking in the wild. Thousands of species have evolved ingenious methods for dispersal, rapid growth, plasticity, and other traits, similar to those used by plants that became successful weeds.

Look at it another way. There are a hundred to a thousand species in the genus Taraxacum. Put them side by side, they all look about the same—a rosette of toothed leaves, characteristic flower, and seeds carried in the wind on a feathery parachute. But only one of them, common dandelion, became a flower of infamy on every continent. Likewise, all hundred or so species of Setaria produce seeds with variable dormancy in a bristly panicle. But there are only a handful, not one hundred, weedy foxtails. The story is similar for nearly every weed of economic significance. How is it that so many were called but so few chosen?

The answer lies in the deeper nature of our relationship with weeds. Plant evolution is an indifferent process that operates through chemical changes at the DNA level. It's all about variation and selection, a careful biological form of choosing. But the making of weeds from the dawn of the great Revolution through modern adaptation to lawns, gardens, fields, and waste places has not been simply detached and unemotional. Our connection with ill-mannered plants is deep. Most of the important weeds began their association with people as plants of value for food, ornament, medicine, or other useful material. They might have simply been objects of curiosity, colorful flowers, twining stems, seed capsules that rattled in the wind. People touched them, tasted them, fed them to animals, harvested, stored, transported, replanted, tended, and nurtured them. And weeded them. They thrived along with the people who enjoyed them and developed emotional attachment to their beauty, aroma, or taste.

At some point, disillusionment set in. Expectations weren't met. Humans found other species more attractive, tasty, or compliant. The relationship went sour. But weeds had already evolved structures and physiology to function in the world of human activity. The more humans pushed them away, the more the uninvited undergrowth evolved ways to stick around. Not all of the paths to weediness involve such unrequited love, but all the herbaceous characters in these stories found a place in the human heart and mind that goes beyond mere biology.

Agriculture, after all, is a public act. Fields, gardens, and lawns are always on display. So weeds took on symbolic meaning connected to public norms, cultural values, and attitudes. Agriculture is also a major economic activity that inevitably intersects with limited and shared natural resources that make it function. Pestiferous plants became targets of leading corporate interests; keeping weeds under control helped farmers, pleased the shareholders, and kept food prices low. At some level, green leafy trespassers and their control became part of everyone's life. Not just the pesky things in the garden, flowerbed, grain field, and vacant lot. They're in the cost of everything that comes from a farm, every "non-GMO" label in the grocery store, allergy-inducing pollen in the air, pesticide drift from your neighbor's lawn, the curled leaves of your fruit trees following a herbicide sprayed a mile away.

Now, as we find ourselves firmly in the Anthropocene, weeds and their impacts are no longer confined to gardens and fields. They spread and share genes at a global scale. The widespread dispersal of plant propagules to human-disturbed habitats across the planet has collided with major disruptions of biogeochemical interactions that compromise the long-term sustainability of ecosystems on which all life depends. Directly or indirectly, this threatens the survival of many species, including humans.

As weeds evolve in this anthropogenically disrupted environment, agrestal selection is at play, shaping the outcome. But now the selective forces are beyond nature and even beyond agriculture; they are global selective pressures of the Anthropocene where species like ragweeds threaten agriculture, endanger human health, and thrive on pollution, climate chaos, and environmental dysfunction.

The Deeper Essence

Weeds come, and weeds go. Not only do the species change, but the view of what is, or is not, a weed continues to change over time and place. Neither weeds nor weediness are static. No wonder there is no fixed botanical definition of "weed" or timeless agreement on which plants merit such derision. Humans have helped to make thorny thugs and to unmake them. Plants that any farmer or gardener would recognize as a pest today were not always so. Most of the species discussed in these chapters were at one time considered desirable, or at least innocuous. Some were coveted, protected, and enjoyed for ornamental, medicinal, culinary, or other practical purposes. Conversely, the thorniest enemies of farmers of old include some species that are insignificant today; some can hardly be found.

My favorite example comes from William Darlington's 1847 botany book in which he itemized the most serious weeds of North American agriculture. His list included oxeye daisy, star-of-Bethlehem, and bracted plantain.[9] Similarly, *The First Ohio Weed Book*, written by A. D. Selby, ranked red sorrel among the most troublesome plants facing Ohio farmers in 1897.[10] I've presented these lists to many farmers and gardeners. Most have never heard of these weeds or regarded them as problems.

The reason is simple. Farm fields of the 1800s went without lime, and soils were acid. Crop seeds of that era were contaminated with seeds of acid-tolerant daisies, plantains, red sorrel, and others. Modern gardeners and farmers, with mechanization, better managed soils, and efficient seed cleaning have other weeds to contend with, those they've inadvertently favored by modern means. The weeds that American farmers consider most troublesome today include marestail, waterhemp, and Palmer amaranth.[11] Neither Darlington nor Selby even mentioned these plants as objectionable. In fact, these species weren't on any list of troublesome weeds ten years ago.

Weeds come and go across cultures as well as in time. I recently asked an Amish friend which species were most bothersome to him. He is an excellent farmer with fine crops of hay, grains, and vegetables. I drove past fields infested with ragweeds, pigweeds, and foxtails to get to his farm, which operates with draft horses and practices dating to the 1800s. He explained that throughout the whole Amish community in the hills of eastern Ohio, the most vexing plant is yellow buttercup. It grows in horse pastures. No matter how low you cut them, they survive and compete with the grasses and clovers that sustain the main source of power for the farm—horse power. Foxtails and pigweeds, he explained, follow compaction by heavy equipment and are not problems on Amish farms, where horses do the work.

These examples explain why weeds are social, cultural, psychological phenomena as much as they are biological wonders. Most efforts to "control" unwanted vegetation rely on technological solutions that are blind to this deeper essence. The result is ultimately frustration, which works to the weeds' benefit. The deeper essence of weeds is revealed in the attitudes and behaviors of humans, which are the source of the frustration. This does not minimize the serious economic and practical problems that noxious grasses and broadleaves cause for farmers of every crop throughout the world. Understanding weeds in all their dimensions, and appreciating their biocultural histories, illustrates how certain plants became especially troublesome and how changes in attitudes and behaviors might help make them less so.

Down Eight Weedy Paths

The eight paths to weediness that I'll describe represent just a few of the coevolutionary trails that humans and unwanted leafy companions have traveled. They all show how agrestal selection has driven changes in plant populations, usually for the benefit of the weeds. They also show how humans, with mostly good intentions, have modified undesirable flora in more than a biological sense and given them meaning that ordinary, inconsequential plants lack. The trajectory of the stories from the suburban world of dandelion to the global impact of ragweeds is one of increasing severity and impact on human health and the environment. I end with foxtail, from a group of grasses that can be weeds or crops or wild or waifs, representing the hopeful imagining of different possibilities and choices for how plants can be, and maybe humans as well.

Ultimately, this book is about our role in the natural history of weeds. I've attempted to give appropriate attention to the behavior and consequences of the other half of what the word *weed* signifies. Humans and weeds have been ecologically successful because of our remarkable ability to adapt to conditions around the world. The ability to survive in a wide range of environments comes from a long history of accumulating genetic variation and, specifically, genes for plasticity. Humans, with our big brains that give us the ability to use symbols, explore, and create, can harness our collective good will to sustain the planet for the happiness of future generations. Or we can ignore what we're capable of and rely on our baser—weedier—instincts to simply grab more for ourselves.

Inevitably, natural histories that intersect with human history will have winners and losers, conflicts, and uncertainties. Natural histories don't offer solutions. They are an accounting of how we got where we are today. These chapters tell how we got the weeds that farmers and gardeners face now, why pest plants persist despite increasingly sophisticated chemical and genetic technologies aimed against them, and how all of us are part of that. Going forward, we can learn from this history. Or not. Learning from it means all the things reasonable people already know about how to live on this planet in ways that are more respectful of the environment, the biosphere, and each other. Because weeds are known to everybody, the stories of personal interplay with humble botanical accomplices are a way to explore consequences of modern life that might not be appreciated. My hope is that these narratives will stimulate interest in evolutionary biology and provide some basic language to help make the science more approachable so that better understanding will lead to better decisions about how to live, how to eat, how to vote, how to appreciate and enjoy the natural world.

Dandelion, *Taraxacum officinale*.

1

Dandelion

In the late 1980s I moved back to my boyhood state of Ohio to take a job doing research and outreach education on weedy plants. Ohio was an ancient crossroad for travelers who spread a diverse mix of weed species along its rivers and trails. The promise those travelers sought in the rolling hills and verdant plains had been tarnished by industrial and agricultural contamination of Great Lakes and Mississippi River watersheds. The environmental movement had started, in part, in response to industrial fires on the Cuyahoga River, on whose watershed I was raised. Now, weed-killing herbicides were polluting the air and water. I'm from the *Silent Spring* generation; I felt compelled to be part of the movement that would find safer ways to grow food and other crops. That's why I had studied weeds in the first place.

Word got around about the new weeds guy in town, and I started getting phone calls and visitors. Farmers drove for hours to share struggles with weeds, just like the specimen they'd kept under the car seat for three weeks. Minifarm homesteaders presented pollen-puffing bouquets with a look that begged for deliverance. I enjoyed the puzzle of new weeds spreading into wheat fields and stories of botanical blemishes "taking over" homeowners' yards and lives. It was a way to put my training into practice, using weedy plants to teach about ecology and the environment.

From spring to fall came questions about common dandelion (*Taraxacum officinale*). With its butter yellow flower and whiskery seed "puffs" that delight children everywhere, dandelion is among the most easily recognized plants on earth. It is also one of the most deeply despised. Between delight and antipathy were things about dandelions that didn't make sense to me.

A bush in the forest, a roadside bloom, a spring ephemeral, a pasture wildflower—people pass them by. They trigger no emotion (maybe joy), evoke no response (maybe serenity). Not so the dandelion. Why the difference? The plant is not competitive, aggressive, or overtly obnoxious. It has no barbs, spines, or nasty odor. It is neither poisonous nor ugly. I had worked with serious weeds that destroyed fields, poisonous weeds that killed livestock, and parasitic weeds that sucked dry the crops that would feed a family. But dandelion simply filled little gaps in a lawn. I could hardly call it a weed.

When sincere, anxious homeowners brought me a specimen, I couldn't help asking if the offending plant was doing real harm, making anyone sick, interfering with performance of daily functions. After an uncomfortable pause, the response was never, "Help us find a way to live with it." Instead it was, "Look, just tell me what to spray so I can kill it." Somehow, that yellow flower surrounded by a flat rosette of leaves had become enmeshed in a sense of personal status and moral decency. People willingly spent scarce resources and risked their own health and safety to be rid of it. I saw an insignificant yellow-flowered stray. Others saw something else. As the weeds guy, I had to figure this out.

The End of Sex

Dandelions belong to a huge plant family, the Asteraceae, which includes sunflowers, chrysanthemums, and about twenty-three thousand other species. The Aster family evolved between 60 and 30 million years ago on the

Gondwanan land mass at the south of the globe.[1] They drifted with wind, water, and whole continents across much of the planet. Along the way they perfected a remarkably successful reproductive system, a composite flower comprised of many small tubular florets packed on the end of a thickened stem. Pollinators easily find a mass of florets—sometimes hundreds—to pollinate quickly. Successful pollination ensures healthy seeds that can spread new generations of plants. One group of Asters became isolated in remote valleys of the Himalayas where, after a few million years, the *Taraxacum* genus arose. These were the ancestors of common dandelion.[2]

The dandelion ancestors were sexually promiscuous. Pollen flew from male anthers to female stigmas of flowers on the same plant and flowers of their neighbors. Genes from one plant got mixed up with those of others in millions of new combinations. Every combination of genes (genotype) was a genetic experiment, a test of traits for survival and for making more would-be dandelions.

One successful gene combination produced parachutelike appendages that carry seeds into the air. This feathery structure, known as the pappus, folds itself in, allowing many seeds to be packed on a composite flower head. It folds out to carry seeds away on updrafts of air. The pappus is aerodynamically engineered so a seed falls, not straight down, but at an angle, and not any angle, but the angle that allows tiny barbs on the seed edges to anchor the seed in the soil. Another successful genetic experiment led to a hollow stemlike scape, a flexible stalk that won't break in a strong wind. The scape shoots up to display a blooming flower head, bends down to protect developing embryos, and shoots up again to disperse the matured seeds.

Protodandelion's millions of years of licentiousness yielded more than two thousand *Taraxacum* species. Some of them evolved red seeds, others black; some made toothed leaves, others lobed. Some adapted to survive very cold conditions; others to survive deserts. It seemed like this shamelessly prolific mating and mixing of genes, which botanists call outcrossing sexuality, flourished among species that moved, adapted, and reproduced in a wide range of environments.

After 20 million years of evolution with wanton sexual mixing and selection of the fittest gene combinations, things got complicated for ancient dandelions. Some of the species essentially gave up sex. They had enjoyed one of the most elegant reproductive systems in the botanical world. Then they tossed it aside. Nobody knows why. With a few mutations, they nearly—though not totally—put an end to genetic recombination and experimentation. What they adopted instead was a mode of reproduction called apomixis.

To understand apomixis, it helps to consider the normal pattern of plant sexual reproduction. It starts when a parent cell, with two copies of each chromosome, undergoes meiosis. Chromosomes pair up and break apart. Pieces of DNA move around, cross over each other, and are reassembled in a new order. The result of meiosis is four daughter cells. Most important, bits of DNA get mixed up, turned around, and reshuffled as a new genetic experiment. This recombination of DNA during meiosis provides new combinations of genes that code for new sets of traits. This is the source of genetic diversity in plants. It provides the variation on which the forces of evolution can act to select plants with combinations of genes—and thus traits—that survive and reproduce successfully in new environments.[3]

But in apomictic dandelions, meiosis comes to a halt before chromosomes pair up and exchange DNA. This meiosis interruptus is the paradoxical nonevent of apomixis. Apomictic dandelions have no chance to make novel gene combinations. They bypass the messy business of fertilization, and flowers develop directly into embryos with the very same genes as the mother. The seeds they send into the air are essentially clones.

No meiosis, no rearrangement of genes, and no novel genotypes mean no hope for adaptation to diverse environments across the world. No way to survive the next climate change. From the standpoint of evolution, apomixis in dandelions is a dead end.

Like all plants, dandelions live in a world of struggle for resources—sunlight, water, and a handful of essential nutrients. Ecological success requires evolutionary adaptation to maximize fitness in diverse environments where dandelions can grab resources from across the globe to make even more dandelions. But apomictic stubbornness stifles any hope for genetic advancement. A crippling mutation would be passed on to identically crippled dandelion progeny. With more mutations over time, dandelion cubs would be less healthy, less able to survive stressful conditions.[4]

Even without the hand of genetic recombination tied behind their back, ancient dandelions stood poised for insignificance. Their survival strategy was to hope some seeds fall onto a small scratch in the earth not occupied by another plant. The small seedlings grew slowly and never extended their leaves high enough to shade other plants. They hunkered beneath brushy, grassy plants that sprang up tall and fast, sucked away nutrients, and captured sunlight at their expense. As simple perennials, they had no creeping roots or stems to spread, occupy more space, or seek more favorable sites. Flat leaves without spines, thorns, or prickles were so weakly toothed as to be laughable among potential predators. The only defense they evolved was a sour milky latex in older leaves and roots.

Yet apomictic dandelions survived the trash heap of evolution. Lucky for them, the apomictic lineage had started with genetically variable plants, including some that sneak off to illicit sexual recombination encounters. This low frequency of recombination produced enough genetic variation, and over time other random mutations occurred due to environmental stress. In spite of themselves, dandelions became one of the most widespread plants on the planet. Lucky for them, a path to weedy success was opened by a curious upright, bipedal hominid with opposable thumbs and a big brain, who happened to roam the landscape as the glaciers receded.

Toothy Attraction and Utility

Early humans could not help being attracted to—and I suspect delighted by—dandelion's shiny yellow floral display. How could they not have reached down to feel the soft inflorescence or have been amused by the grey-bearded globes that sent forth seeds that float in the air? Certainly, they would have tasted the young leaves and flowers. Humans had wandered to almost every land mass by the end of the Pleistocene Epoch (more or less twelve thousand years ago). By this time, dandelion ancestors had crossed the Bering land bridge.[5] The hybrid swarm of dandelions didn't need to rely on a long hopscotch of seeds from Asia into what became North America. More likely, dandelion charisma convinced humans to facilitate their dispersal, by accident or on purpose, everywhere people traveled, gathered plants, and interacted with the natural world.

Unlike other plants that commanded human attention, such as wheat, beans, and potatoes, dandelions did not invest in large or tasty structures that store sugars and protein. Doing so would have followed the rules of the crop world. Instead, dandelions tossed a few meager resources toward humans. The young leaves are a palatable, though bitter, source of vitamins and minerals. The flowers are edible but not especially tasty. Over time, dandelions became known as a diuretic and a laxative. This modicum of food and medicine was enough to convince humans to prepare the ground for dandelions and clear the land of competitors, so they could flourish and release yet more seeds into the air. They were a vagabond species that danced in the footsteps of human migration, settling down wherever humans scratched the earth.

Ancient Chinese healers used dandelion roots and leaves in herbal tonics. The Greek goddess Hecate, overseer of the earth, moon, and underworld, allegedly prepared dandelion salad greens. Early Romans collected and tended dandelions for food and medicine. Across Europe, dandelion leaves were mixed in stews. Flower heads were battered and fried; the Celts made them into wine. Roots dried and ground made a tonic and decaffeinated drink. Dandelion gained a place in physic gardens of Medieval monasteries and hospitals. Monks and peasants prepared soil and planted dandelion seeds, tending them, weeding them, protecting them from rabbits and deer, and praising their maker.[6]

The importance of dandelions in human cultures is revealed in languages of the world.[7] The American common name, and similar versions in Romance languages, is a corruption of lion's tooth, *dent de leon*, which dates back to the thirteenth century. This could refer to the shape of the developing seed, the root, the indentation of the leaves, or the shape of the ray flowers. Theophrastus, the ancient father of botany, referred to it as *aphake*, possibly indicating its use in springtime before lentils (*phake*) appear. Pliny the Elder gave it the name *ambubeia* from the word *ambulo*, going back and forth, suggesting the scattering of seeds. One German version is *Pusteblume* (breath flower), from seeds blown in the air. Various versions of the French *pissenlit* (pee in bed) refer to the diuretic effect of the plant, based on claims that it opens the urinary passages. To the

Italians it is *girasole dei prati* (meadow sunflower). To the Chinese it is "earth nail" (*púgōngyīng*), a reference to its substantial taproot.[8]

New World Salad

When it seemed like all the sites for dandelion habitation in Eurasia were occupied, people from Western Europe obliged by making their way to North America. During the age of colonization, early settlers carried dandelions as a pot herb, for food and medicine. There is some dispute about dandelions riding on the Mayflower, if not purposefully in apron pockets, likely unintentionally in animal feed, fleece, or digestive tracts.[9] Regardless, that would have been a relatively late introduction to the continent. Indigenous people from the plains already knew of it, probably from lineages that had crossed the ancient land bridge. Genetic variation was assured by multiple introductions of dandelions by successive waves of settlers from Holland, Germany, France, and England. Arriving Europeans planted and tended dandelions in gardens up and down the Atlantic coast. Swedish explorer Peter Kalm, writing of his trip to New England in the early 1750s, reported that "common dandelion . . . grew in abundance in the pastures and on the borders of the grain fields. . . . In the spring when the young leaves begin to come up, the French dig up the plants, take their roots . . . and prepare them in vinegar as a common salad."[10]

Dandelion followed Europeans' westward movement across the continent. Plants and people colonized the landscape in tandem. Where the forest grew too tall and shady to allow dandelions to grow, settlers cleared the way by chopping down stands of inconvenient virgin woodland to provide suitable habitat for parachuting seeds to fall. Where grasses and forbs grew so thick that dandelions could not get established, pioneers eagerly burned down the native prairie and rutted the land with their horses and wagons to prepare sites for dandelion seedlings.

Through its charming appearance, meager food value, and modest herbal utility, dandelion spread across the New World. Americans prized it for its durability, relied on its persistence, and enjoyed its cheerful attractiveness. In return, dandelions were planted, cultivated, and nurtured.

By the 1800s, it was a common crop in home gardens and fresh markets. New Englanders selected and named specific varieties; they were sold through seed catalogues.

Home gardens, of course, are a particular thing. They are personal, often hidden in the back, and if a little untidy, well that's just how a garden is. Known in her lifetime more as a gardener than a poet, Emily Dickinson abided the dandelion and beginning in the 1850s, wrote four tributes to it.[11] In some primal and innocent way, through her expression, it was loved. Yet for dandelion to be truly successful as a species, to ensure its future spread and survival, to achieve the highest level of ecological fitness, it remained for the dandelion to be truly despised.

Into the Heartland

Growing up in Ohio in the 1960s, I gave little attention to yellow flowers in the grass around the house. Our yard, carved out of exhausted clay loam soil, had a mix of them, along with white and pink clovers, purple violets, and unremarkable grasses that made a great place to roll around or toss a ball. But the lawn across the street was a green carpet. One summer day, my mother issued us kids various trowels and forked implements, with instructions to rid the front yard of dandelions. I don't know what motivated this. She was not a keep-up-with-the-neighbors sort of person. She was, however, pretty good at prodding six kids out of the house. Anyway, we soon realized it was hopeless. We'd never be able to dig them all up and went inside to try to wash the sticky brown stain off our hands. Besides, not even Mom could come up with a reason the dandelions shouldn't be there.

We could not have known that the idea that our little brick house should sit behind an open patch of mown grassy plants—a "lawn"—was rooted in the well-tended estates of the English gentry. Thomas Jefferson was so taken with the image projected by these estates that he had a lawn installed at the University of Virginia.[12] The father of American liberty drafted enslaved Africans to level the soil and sow the seeds. Soon thereafter, the American concept of the large, open expanse of green lawn came to be associated with wealth, property, moral values, even gender roles.

In Jefferson's time, lawns held a mix of clover, grasses, and forbs, a diversity of species that was considered desirable.

A stately image could not be projected if weeds grew too high. And dandelions could not establish well if grass was too thick or tall. Dandelions were opportunists that took advantage of small gaps where nothing else grew, patches where grass had been trampled, or where the scythe cut too low. Lawns, stately images, and dandelions came together in the Machine Age when cotton gins scattered across the South while mowing machines—and dandelion seeds—scattered everywhere else. The first heavy reel-type mowers appeared around 1830. For the mowers to work, the lawn had to be kept short. People strong enough to push the darn things spent hours cutting the lawn in the heat of the summer. This removed any shade from taller grasses and forbs and gave dandelions more sunlight and greater freedom to spread their seeds.

After the madness of Civil War, the American continent staggered in disarray. It was impossible to repair lives or control the sense of chaos. But maybe the grass, the disorderly meadowlike openings, could submit to the need for order. And thus the clipped and uniformly green expanse of lawn became a mandatory fixture of the American family dwelling.[13] Over time, the notion of the idealized manicured lawn, mossy green, lush, short, and neatly edged like a crew-cut soldier standing at attention, led Americans to rethink their toleration of yellow-flowered interruptions. Grey-bearded puffballs that dispersed in the wind lost any allure they might have had. Emily Dickinson's lovely "shouting flower," her cheerful "proclamation of the sun," simply became a reminder that winter's end signaled the drudgery of weed pulling. Longfellow's "maid with yellow tresses . . . changed and covered o'er with whiteness . . . [had] Vanished from his sight forever."[14]

Even without poetic visions, the human imagination would be good to dandelions. Dandelion would inspire Americans to adopt behaviors never before contemplated. Gatherings of mostly women and children turned away from homemaking and community activities to spend long hours on their knees cutting dandelions out of lawns with knives or tools especially made for this purpose. Dandelion took advantage of this practice because it can regenerate from root tissue below the crown. Cutting a single dandelion crown merely stimulated growth of buds deeper on the root. Where once there was one rosette of rough-toothed leaves, now there were two or more. Toolmakers invented special rakes to strip the flowers off before they went to seed. Crews of workers moved back and forth across the lawn in the mid-day sun to pop the tops off the stalks. This left a patch of tall rubbery dandelion scapes, whose ragged form could hardly have been an improvement over the flowers themselves.

Dandelion responded to human manipulation with the tricks available to an apomictic species. It made the most of the low level of genetic variation it had acquired in the past. The genotypes best suited to the lawn habitat survived and reproduced. As apomicts, these lawn-adapted plants parachuted their clonelike seeds into the next available lawn. Dandelion also made use of plasticity, the inherent ability of individual plants to modify growth in response to their surroundings. When dandelion plants were cut low with those special rakes, plants responded by making shorter scapes, to keep their flowers closer to the ground. This avoided the reach of the rake and saved the energy needed to make taller scapes. Plants used the saved energy to make more flowers, seeds, and roots.

Dandelion ancestors had spent a few million years evolving the pappus for seed dispersal. But it only took a few decades for finicky belt-driven reel mowers to be supplanted by a rotary version that could shoot dandelion seeds across the lawn. Rotary lawn mowers with a lightweight engine connected directly to a rotating blade replaced evolution's random parachute approach with human-aided seed dispersal that was significantly more uniform. Whole neighborhoods of low-cut lawn expanded, and dandelions found the space to stretch out.

Dandelions exploited every human manipulation of the environment. Humans cut down trees, cleared land for look-alike housing allotments, and tamed the grass. This made more sunlight and nutrients available to dandelion seedlings. Dandelion photosynthesis converted sun's energy into extra leaves, deeper roots, and more seeds. A subset of humans, with their own highly advanced plasticity, became lawn neat-nicks, obsessed with mowing dandelions into oblivion by cutting closer and more frequently. The highly plastic dandelions simply laid their leaves flat against the soil surface to escape the blade. Even the neatest neat-nick can't keep up this cutting regime forever, and during short intervals between mowings, dandelions replaced leaves rapidly. When the clunky old lawn mower was in the shop and grass grew taller, dandelions pushed their leaves up in all directions to form a moundlike canopy, allowing sunlight to reach the leaves more effectively from all angles.[15]

People had been burning vegetation to suppress weeds for centuries. It's no surprise somebody would get the hot idea to do so on behalf of lawn purity. Plus, it's a chance to use a flame burner for a respectable purpose. You simply lug a tank of highly explosive gas around and put the torch to every dandelion rosette. One at a time. Of course, dandelions mostly sniff at this treatment. After a few leaves go up in vapors, dandelions simply send up new ones from crown buds below the soil surface. Nevertheless, this approach provides the satisfaction of watching weeds crisp up and gasp at the mercy of a scorching tool.

Back on the farm, dandelions were confined to poor pastures and waste areas. Horses gave way to gasoline-powered machines beginning around 1910. Tractors prepared the ground for the spread of dandelions into virtually every cropping system in North America. The area of tilled soil expanded dramatically, and the mixture of crops and animals was ideal for expansion of dandelion populations. After three years of growth in a hay or pasture field, dandelion infestations grew so large and produced so many seeds that it was nearly impossible to control them when the field was rotated to corn.

New farming practices that reduced tillage to conserve soil were developed in the 1960s. They created an environment for dandelions as stable as a lawn, but better. With less tillage, dandelion roots grew deeper. Seeds found safe places for germination among crop residues. Farmers controlled other weeds, applied fertilizer yearly, and harvested crops just in time to give dandelions access to sunlight during its peak growth periods in spring and fall. Environmental conditions during autumn are especially good for dandelions. The sunny days and cool temperatures allow for high rates of photosynthesis to build up energy reserves in the roots. This ensures winter survival and vigorous spring growth.

Eventually, dandelions became a significant member of the weed flora, not only in lawns, but also fields of corn, soybean, wheat, and alfalfa. And today, in the springtime, across the Midwest, you can spot a field of clover or alfalfa from miles away. They're infested with a mass of yellow, and the air is filled with seed-bearing parachutes that proclaim a triumph of ecological fitness.

Securing a Dandelion Future

In the last few years of his life, Charles Darwin became preoccupied by the movement and twisting of plants. He wrote a whole book about it, describing the helical climbing of plants like bindweed. Something in the tip of a young seedling, he speculated, causes the shoot to bend toward light.[16] The book was not a big seller, and it remains one of his least known volumes. But in the 1920s, Dutch researchers picked up on Darwin's idea and isolated a chemical, a plant hormone, that regulates plant growth.[17] This was exciting to plant biologists—maybe they could use hormones to make plants grow faster, bigger, more fruitful. Ingenious chemists stabilized the chemical, so it worked even better to modify plant growth.

They called these new chemicals "auxins." Research groups in England and the United States developed one auxin called "2,4-dichlorophenoxyacetic acid." They gave it the uninspiring name 2,4-D (pronounced simply "two-four-D"). Hormones have different effects depending on the concentration applied. At very low doses, 2,4-D could increase plant growth. An early report showed that 2,4-D was three hundred times more effective than other auxins for making seedless tomatoes. The only problem was that if you got the dose a little too high, 2,4-D had a nasty habit of killing the test plants. Without much excitement for dead, seedless tomatoes, interest in auxins withered.

An idea took shape around 1941 that the tendency of 2,4-D to kill plants might be useful. With world war heating up, a plant-killing chemical caught the interest of the military. Further work was conducted in secret at Fort Detrick, where the army's center for biological warfare tested thousands of crop-destroying chemicals. There are no reports that 2,4-D was sprayed on enemy crops in World War II. That would wait for war in Vietnam, when 2,4-D and its cousin 2,4,5-T were spread across jungles, crops, and inevitably people. Around the end of World War II, manufacturers of 2,4-D contemplated possible markets for this chemical.

The world was a dangerous place in August 1945. FDR was dead. Truman was preparing to drop a second bomb. The Potsdam Conference was about to end badly, splitting Europe into zones of influence. Trust between East and West faltered. American and Soviet armies would soon have enough nukes to bomb each other back to the Stone Age. The US troops returning from Europe were filling the suburbs and finding lawns poorly tended. Word got out that a foreign agent had made its way from Russia, down through Canada, and was threatening homes across America.

In a mostly untold story of the Cold War, a young botanist, Fanny-Fern Davis, stepped toward her place in history. Dispatched in a white dress and street shoes, Ms. Davis sprayed 2,4-D to great effect, killing the dandelions infesting the National Mall, and saving the nation from disgrace just blocks from the Capitol.[18] The Soviets might have launched the first intercontinental ballistic missile. They might have been first in space. They might have had sixty-eight thousand nuclear warheads pointed our direction. But America had 2,4-D. And American capitalism would secure American freedom from yellow-flowered weeds.

The nation united to respond to the threat of capital D Dandelion.[19] Whole industries mobilized to design, manufacture, market, deliver, repair, and insure an incredible range of consumer products whose main purpose was to make American suburbs safe for conformity. Inventions that had been developed for the farm were miniaturized for the suburban half acre of green. 2,4-D and other herbicides that farmers sprayed to kill dandelions in corn fields were repackaged with macho names and jacked-up prices for sale across urban America. Millions of three-gallon sprayers were sold to dispense novel toxicants in every village, town, park, and roadside.

The campaign against Dandelion built a new international agricultural chemical industry. Teams of Ph.D. chemists worked to synthesize new complex molecules to disrupt biochemical pathways in dandelion. Crews of laboratory, greenhouse, and field technicians screened hundreds of chemicals every week to find one with promise. By one estimate, more than 150,000 chemicals were tested to get one to market, at a cost of more than $250 million.[20] Between 1950 and 1970, 2,4-D became the most widely used pesticide on the planet.[21] Even into the 1990s, an industry representative told me: "Dandelions are fundamental; we wouldn't waste time" with a chemical that did not attack this weed.

To get some perspective on this, I contacted the Scotts Company, a major supplier of lawn-care products based in Marysville, Ohio. A friendly email responded that they marketed about half a billion dollars in lawn herbicides; three-quarters of that against Dandelion, "our primary advertising focus."

With the rise of 2,4-D, the future for Dandelion was assured. This might seem counterintuitive, because 2,4-D does, indeed, kill dandelion plants when applied at the right growth stage and dosage. This is the paradox of weed control. Every action to remove dandelions from a particular habitat creates a small disturbance, a gap that a new dandelion seedling has the opportunity to fill. The reward for pulling, cutting, and burning is brief because dandelions readily regrow from buds near the top of the root. Initially, hope abounded that 2,4-D would at last put a halt to Dandelion. A quick glance out the window in springtime shows how well that has worked out.

Calibrating an Entanglement

The Dandelion imperative found me sitting on the cold floor of a garage on a beautiful Ohio spring morning in 1969. Summers and weekends during high school, I was lucky to have a job doing yard work for a patient and generous couple who somehow overlooked my tendency to lose tools and break equipment. In front of me that morning was a bag of something called "Weed-n-Feed," an ingenious product with a colorful label that promised to kill weeds and feed the grass in one giant step for all lawnkind. What could be better than that? I gave up a brief struggle with the raveled drawstring closure and pulled at the sides of the bag. Five pounds of pesticide and fertilizer dumped out in front of me in a big poof of smelly grey dust.

My instructions were to calibrate the spreader and apply the product to the lawn all around the house, "and be sure to read the label first." So, I scooped what I could back into what remained of the bag, put some into the green and yellow spreader, and the rest I swept into the patch of oil that had dripped beneath the car. I tried to reassemble pieces of the torn bag to read the label, but the print was small and full of unfamiliar words and long chemical names that meant nothing to me. "Harmful if swallowed." Okay, I got that. No word about breathing in a cloud of dust. Spreader calibration instructions were impossible to understand. Not being especially good at math anyway, I decided to just open the slot on the spreader a little bit, to make sure I didn't apply too much and kill the whole lawn. But if I didn't put enough on and the weeds didn't die, it would look like I hadn't done the job. After all, the dandelions were already blooming, and they had to be killed. I don't recall how I resolved this dilemma. I do remember making a second and third application over the next few weeks.

Two decades later, I was back in Ohio as the weeds guy. I moved with my family to an old farmhouse on the edge of town. After a few weeks, the man living next door started to send me rambling, threatening letters because I didn't keep a manicured, dandelion-free lawn. Someone at the county commissioner's office warned of a fine. That's when I began to see a deeper dimension to the impact this species has on human attitudes and behaviors. I was no longer an observer. I was entangled in the convoluted cross-species choreography between Dandelion and humans.

I know that biophobia is a real thing: I remember this every time a snake crosses my path. Visceral reactions don't wait for cerebral machinery to judge whether it's a threat. But the prevailing suburban American reaction to Dandelion is subtle, understated. Yes, the plant can cause a mild dermatitis in some people. It could make you sick if you could stand to eat a lot of it. But scouring medical journals, I have not been able to find a single case where dandelions were the proximal cause of illness in an individual whose exposure to the plant was through the parlor window.

In graduate school, I had studied things like photosynthesis, plant anatomy, and seed germination—the things I thought weeds were made of. None of that training prepared me to deal with whatever was happening here. Nothing in plant science spoke to the origin and nature of the aversion, the fear. I searched the scientific literature and consulted with learned colleagues to find a nugget of insight that might explain the depth of what, in the presence of Dandelion, appeared to be going on. Inevitably, I suppose, I turned to Sigmund Freud.

Yellow Disgrace Full of Grace

One of Freud's earliest (1899) psychoanalytic writings concerned a dandelion memory. It was written obtusely, as a conversation between a narrator and a thirty-something academic, a faintly disguised Freud. In a 1946 translation by Siegfried Bernfeld, the memory related a scene in which "I see a thickly covered, green meadow. . . . In the green are many yellow flowers, evidently the common dandelion." In the memory, Freud and childhood friends are picking dandelions. One young girl "has the prettiest bouquet," which is soon torn from her hands by the young Freud and other boys.[22] She runs off crying and is comforted by a woman who gives the children pieces of dark bread. Recalling the story, Freud wondered why this particular experience was held in his memory: "Should the yellow of the dandelion, which today of course I consider not at all beautiful, have pleased my eyes so much at the time?" He explained, "the yellow of the flower stands out . . . too glaringly . . . almost like a hallucination." It reminded him of pictures that intrigued him at an exhibition and "the derrieres of the painted ladies." He reflected that "the representation of hunger is well done" in the bread image, and the representation of love is "in the yellow of the flowers. . . . To take a flower from a girl means to deflower her." He concluded that the dream "must be content . . . with being accepted as a disguised allusion in a childhood scene."

Freud, of course, was distinctly unqualified to render a cogent analysis of a plant species about which he clearly knew little. My own interpretation of Freud's Dandelion memory is straightforward. The brightness of the yellow, and its connection to the "derrieres of the painted ladies," needs no further elaboration regarding Freud's particular proclivities. More interesting is that dandelion yellow, which so attracted him when he was young, he now "consider[s] not at all beautiful." The overt sensuality explains why it must remain "disguised." In other words, the primal attraction to dandelion was acceptable as a child's infantile sexual curiosity. But when we put childish ways behind us, Dandelion becomes a symbol of paraphilic urges and temptation, an embarrassment to be hidden away with our shame.

This interpretation explains why Dandelion is doubly offensive to those who would rid it from their lawns. Dandelion is a flaw in the orderliness of green uniformity, a mark of slovenliness, a subversive affront to the social order of suburbia. People believe they hate it because their neighbors are thought to hate it, and the social contract of the neighborhood stipulates that everyone will maintain order lest the whole place goes to hell. But at a deeper level, repulsion of dandelion represents a fear of the id, the primitive and infantile.

Walt Whitman, who obviously knew something about lawns, having spent his life writing a whole book about grass, described this in "The First Dandelion."[23] Whitman likened the flower to a child, "simple and fresh and fair." It is "innocent, golden, calm as the dawn." It dares to "show its trustful face" by emerging into an adult world dominated by the "artifice of fashion, business, politics . . ."

Thus, Freud's libidinous Dandelion cannot be tolerated because, like the id, it satisfies childhood delight in beauty and an impulse to fly freely in the wind, uninhibited by adult seriousness, repression, and sense of propriety. This underlies suburbia's proscribed sexual repression, which prohibits hanging laundry to dry, pubic grasses to grow along the edge of the sidewalk, and dandelion flowers to bloom. Get rid of the offensive Dandelion and stifle the outrageous public florid display transpiring right outside your door. Oversized cars shining in the driveway, phallic-sculpted bushes, and taut teenage bodies sunning themselves in the backyard are okay, because they are displays of wealth and leisure. They exalt what the suburban dream is all about. Only poor folk or sinners let dandelions bloom.

What doesn't make sense with respect to dandelions, from a weed ecology standpoint, is that these plants, as we've learned, are apomictic. The dandelions of Freud's yellow-flowered temptations, and Whitman's seductively fresh fairness are completely nonsexual. They're as chaste as the Virgin Mary statue in the side yard.

Dandelion Dealings

Living the suburban dream takes more than high moral values: it also takes money. Dandelion's ultimate success across America would be found in the pocketbooks of American homeowners. Human emotion comes equipped with caution lights, and dandelion flowers flash yellow danger. More dangerous for upward mobility than prurient signals is fear of financial ruin. And nothing threatens worldly status more assuredly than a weedy lawn. Buyers, sellers, realtors, appraisers, and inspectors know that dandelions do not cause any material damage that would result in actual loss of property value. But by group-think agreement, they believe it is so. And so it is. Already mortgaged to the brink, Americans sink over $900 million annually into lawn herbicides, to insure their assets against Dandelion disaster.[24]

Dandelion and the lawn-care pesticide industry have a sweet deal. The curling up and withering away of dandelions rewards customers with great satisfaction. The offensive plant, or one like it, which will be back the next year, reward the industry with future profits. The success of both Dandelion and 2,4-D are aided by regular input of seeds, which parachute in from elsewhere and lie dormant in the soil until conditions are right for germination, and by roots, which regenerate from small fragments. Success is assured by dandelion's inherent plasticity and just enough genetic variation for selection of favorable genotypes. Consider that if 2,4-D did in fact rid the world of Dandelion, there would be little commercial future for the hundreds of brands and formulations on the shelves in stores and shops around the world.

Dandelion has been an innocent bystander, complicit only in its opportunism and persistence. Its flowers are reminders that it's time to get out the sprayer and make a trip to the lawn and garden store. The salespeople will happily sell dandelion killers even at times of the year when applications of the products are ineffective. Dandelion accumulates carbohydrates in the taproot and moves them upward in springtime to support new growth and flowering. For 2,4-D to kill the plant, the chemical must move downward into the taproot, something that does not happen when springtime plants are flowering.[25] Plants injured by springtime applications of 2,4-D usually survive and recover by autumn. If the plant does die, the slowly disintegrating leaves provide shelter for new dandelion seedlings. Like the actions of that clueless sixteen-year-old sitting on the garage floor, suburban lawn tenders often apply 2,4-D at the wrong time and at the wrong rate, so it's mostly ineffective. If anything, this helps to keep Dandelion in the lawn, as 2,4-D suppresses other weedy competitor species that are susceptible to the herbicide in springtime. Predictably, the human response to ineffective control is another trip to the store and buy more of the stuff, or a stronger formulation to be sprayed at a higher rate and more often.

Collateral Chromosome Aberrations

On a recent November day, I spent an afternoon at the LuEsther T. Mertz Library of the New York Botanical Garden, a mammoth columned structure with a glorious fountain out front. The helpful reference staff, in quiet voices, seemed eager to share the resources of one of the best botanical libraries in the world. I scribbled some Latin names on a scrap of paper and in a few minutes, was presented with a manila folder holding loose publications pertaining to *Taraxacum* species. An anonymous article from a 1938 issue of *Science Digest* caught my eye: "Patents, processes, inventions: Killing dandelions." Based on work conducted in Iowa, the article described how "straight-run kerosene with a boiling-point range of 180 to 250 C, and an unsaturated hydrocarbon content of not over 4 percent has shown to be a very effective differential spray for dandelions." It recommended that the "undiluted kerosene" be applied "uniformly over the lawn . . . at a rate of 200 gallons an acre." Dandelions initially turn dark green for a day or so, according to the article, then yellow for a week, and finally begin their demise with a bronzing of the leaves. Other publications from the time endorsed gasoline as equally effective.

It's hard to imagine feelings toward Dandelion so deep that these were considered to be rational practices, regardless how effective they might have seemed. Both kerosene and gasoline are toxic when inhaled, swallowed, or aspirated. They're skin and eye irritants; ingestion can cause convulsions, coma, and death. They catch fire and smell awful. The articles gave no indication of how long to wait after application before the kids can roll around on the grass again. But we can be confident the dandelions disappeared. For a while at least. Until regrowth was initiated from crown buds and seeds in the soil began to germinate.

I left the library to catch the next train back to the city. Turning around to glance past the giant sycamores standing guard over hundreds of thousands of plant species held at this botanical oasis, I took a deep breath. A verdant patch of emerald green stretched from the marble edifice. One single species of grass—not a weed in sight—sparkled in the sun after a light rain.

The June 1929 issue of *Ladies Home Journal* featured an article touting a new way to produce a "dandelion-proof turf." On a page with ads for "Nonspi antiseptic liquid" that "keeps your armpits dry and odorless" and Williamson "prophylactic shoes" was a recommendation to apply 130 pounds of "sulphate of ammonia" per acre of lawn. Of course, it's possible "the powerful burning action . . . may ruin the lawn." And "occasionally a few unusually deep-rooted" dandelions may survive this treatment, for which "an excellent plan is to dip the sharp point of an ice pick into concentrated sulphuric acid and then stab through the center of each dandelion crown . . ." The sulfuric acid should be kept in a "wide-mouthed glass container, such as an ordinary cream bottle." Never mind that sulfuric acid is extremely corrosive and can cause serious burns, eye and respiratory tract irritation, and tissue damage. At least your eyes, if they survive, will behold a dandelion-free lawn. For a while anyway.

Health and safety hazards of acids and petroleum products must have seemed charmingly quaint when 2,4-D came along in the late 1940s. The new herbicide was formulated for easy application at relatively low rates and with remarkable efficacy. Between 1950 and 1970, 2,4-D production in the United States reached about 400 million pounds (~184 million kilograms), about half a pound (0.2 kilograms) per person.[26]

Around 1970, the safety of 2,4-D and other auxin herbicides became a continuing source of controversy. 2,4-D is the weed-killing equivalent of insect-killing DDT. They were introduced around the same time using arcane "D" laden abbreviations, which has led to confusion. Chemically, and in practice, these pesticides belong to very different structural and functional families. DDT is much more toxic to people and other living things, and much more persistent in the environment, and has been banned in the United States (but not many other countries). 2,4-D, cheap and easy to make, is the most widely used pesticide in lawns, roadsides, and field crops throughout America.

The medical literature on pesticide safety is complicated and inconsistent. For every study linking a pesticide to some type of cancer or other disease, there's another study that disputes those findings. Figuring out the truth on this has occupied the attention of many people much smarter than me. Vilification of the agricultural chemical industry for its safety record has already been done well by others. The industry has done sufficient damage to its own credibility through mishaps and ill-considered policies, to which I can add little.[27]

Concerns about the safety of 2,4-D and similar herbicides have been well publicized. But people remain willing to jeopardize their health and the health of their kids, pets, and neighbors because of a yellow flower that poses no risk. Potential acute health effects from exposure to 2,4-D start with the eyes. It is "extremely hazardous in case of eye contact" and "slightly hazardous in case of skin contact." Anyone so unlucky as to experience a "severe over-exposure" can die. Mutagenic effects are "classified possible for human" as well as other mammals. It is also a possible teratogen, which could cause malformation of embryos in humans. It is a possible reproductive system toxin and a known "toxin to kidneys, the nervous system, gastrointestinal tract, cardiovascular system, upper respiratory tract, and eyes."

These quotes are not from some tree-hugging antipesticide propaganda. They come from the Material Safety Data Sheet (MSDS) issued by a company that sells 2,4-D. The MSDS clearly states that the product is not considered a human carcinogen. This might be comforting to people exposed to it, already worried about eye, skin, kidneys, nervous system, respiratory, and gastrointestinal damage. Experts are still debating the carcinogen findings.[28] The question of risk is complicated by the common practice of combining 2,4-D with other herbicides (dicamba and MCPA) that have equal if not greater potential consequences for human health.[29] These chemicals are applied in formulations with undisclosed inert ingredients, including surfactants, solvents, and carriers, some of which are skin and eye irritants at the very least.[30]

Most of the 2,4-D applied to kill Dandelion never reaches a dandelion plant. Dandelions are scattered around the lawn, so most of the active ingredient falls on other plants or on the soil or pavement or kid's toys.[31] People can pick it up on their shoes and carry it in the house. Kids and pets roll in it on the lawn. Grasses are not killed by 2,4-D because they produce enzymes that break the chemical apart and detoxify it. But it's toxic to honeybees and birds that encounter 2,4-D in the air as a volatile compound or on leaves and flowers and soil. Some applied 2,4-D moves into ponds, lakes, and streams where it can be very toxic to fish.[32] Some evaporates into the air, moves with the wind, and finds its way to the neighbor's tomato plants. It's collateral damage tolerated in the fight for uniformity.[33]

Pesticides like 2,4-D are commonly sprayed on lawns by young men who jump from a green truck and scurry across the yard with a 5-nozzle spray boom. They are the ones most at risk of getting 2,4-D on their skin, inhaling it, or eating it by lunching with unwashed hands.[34] The human body has a remarkable ability to handle environmental toxins, and a dose of 2,4-D moves quickly in and out, excreted in urine, bile, and feces within a few days. Only a small percentage is retained by hooking up with sugars or amino acids.

A measure of the ecological coevolutionary success of Dandelion is its impact on the ecological success of humans. The National Institute for Occupational Safety and Health reports that men who work with 2,4-D are at risk for fertility problems caused by changes in the shape of sperm cells and therefore their ability to swim (the sperm cells, not the men) and fertilize an egg. Research and review articles report that 2,4-D acts as a reproductive toxicant. Men who spray 2,4-D repeatedly have reduced sperm concentration.[35] Studies have reported chromosome aberrations in human cell cultures and increased frequency of mutations in cell cultures of other mammals.[36] Not quite the end of sex, but for some, the end of reproduction.

Contradictions without Contraindications

What am I to make of this as a weeds guy? How can threats to the environment and personal health counter whatever drives the perceived need to exterminate a mostly harmless plant? As an educator, I teach environmental science. But science has no hold over social pressure. People who come to me with weed anxiety express discomfort in spraying poisons over their yard. They dislike putting themselves, their family, and neighbors at risk. Yet they uphold a sense of personal integrity, a community moral code, and a notion of righteousness, all of which are violated by yellow dandelion in a green suburban sea. The head says no, but the heart sees yellow—danger. Greenness just feels right. The tragedy of the suburban commons is complicated by uncertainty as to which shared resource to sustain—air free of pesticides or air free of pappuses.

For weeds, there's always promise in new perspectives. Today, Dandelion represents health. So say labels on products at the health-food store, herbal remedies for diabetes, liver disorders, and as a laxative and tonic. Dandelion leaf juice can treat skin diseases, loss of appetite, and stimulate the flow of bile. The long-recognized nutritional value comes from beta-carotene, fiber, various nutrients, thiamine, and riboflavin as well as vitamins C and D. The local organic co-op in my town sells Organic Dandelion Leaf Powder—using plants from Croatia—just $14.99 for an eight-ounce package. There's also dandelion root powder for a dollar more. A 3.5-ounce bottle of Dandy Blend Instant Herbal Beverage with Dandelion sells for $12.93. Thirty-six Republic of Tea bags made of organic, non-GMO dandelion are only $13.00. Nature's Way dandelion root capsules are $9.99 for a hundred pills, enough to last about sixteen days. The product labels say it is "one of the best blood purifiers" that "restores and balances the blood." In addition, it "helps build energy and endurance" and cleans the liver by increasing the "flow of bile into the intestines" while increasing "activity of the pancreas and the spleen." I didn't even know my liver needed to be cleaned or my spleen activity increased. But that's not all. It also opens urinary passages and is "good for the female organs."

While one industry is focused on getting rid of Dandelion, a growing community of people are working to grow it, protect it (mostly from chemical herbicides and weeds), and harvest it for food, wine, and medicine. Others let it flourish in "natural lawns" for its beauty and pollinator support. Some for the cheeky pleasure of challenging social norms. Thus, Dandelion has harnessed popular culture to tap into the sensibility of consumers living in downtown lofts far from the hum of lawn mowers. There is an interplay between herbal dandelions that are purposefully cultivated—or at least protected—and those in waste areas, roadsides, hayfields, and pastures. These vagabond plants don't threaten the sensibility of lawnophiles. They serve as reservoirs of seeds and enough genetic variation to supply empty niches in lawns across the suburban status-sphere. Now we can hire a lawn-care service to kill dandelions using chemicals that might threaten our health, and then visit the natural-foods co-op to buy dandelion tea, pills, or drinks to clean the toxins from our liver as part of a regimen to maintain our well-being.

Home in the Human Consciousness

More and more, Dandelion and people seem to function in a coevolutionary relationship, each responding to changes in the other to enhance the success of both. The difference is that the motives operate in reverse, as people do all they can to get rid of their coevolutionary partner while Dandelion finds new ways to keep the music humming so the dance can continue.

I've spent much of my career trying to figure out how this two-step has led to the divergent amity and antipathy toward Dandelion. The very heart of the concept of "weed" connects the evolutionary biology of this species with whatever drives human responses to it. In an agricultural context, a competitive, noxious plant that makes food and fiber production impossible becomes a weed. In a suburban American context of social anxiety and image obsession, a yellow flower in a lawn becomes something else, even beside a garden of yellow daffodils or chrysanthemums. In the botanical world, plants in any setting are created equal. Because of their intimate connections with people, Dandelion is not just a plant, and certainly not just a "plant out of place."[37] Dandelion is a social, cultural, and economic phenomenon with different levels of meaning, and its rightful place seems to be somewhere in the human consciousness.

No wonder office visitors and callers were confused when I asked, "Is this weed doing any harm?" To me, *T. officinale* was a botanical wonder, an evolutionary misfit that happened to be good at finding its way into open spaces. I thought my (clever) question might inspire reflection on what is important in life, evaluation of family resources, health, and their environment in relation to lawn purity. But social rules prevail. The offending plant could jeopardize their financial security and that of the neighbors. My visitors were desperate for a remedy. I was the weeds guy. Surely, I should have the answer. They feared for their home equity and their standing in the community. The neighbors, the mortgage companies, the tax appraisers, the lawn and garden store—everybody understood this. Except, maybe, me.

Dandelion might have been satisfied having moved from isolated rugged mountains to more favorable and uniform environments where it has established stable populations. That level of ecological success has not been achieved by hundreds of other *Taraxacum* species, many hardly distinguishable from common dandelion. And Dandelion was probably content to attach itself to the massive human migration from Asia and Europe to North America to simply find new habitat. Its status as a weed was not inevitable. Given less attention, it might have been used as a crop and then abandoned for more promising species.

Instead, Dandelion's path to weediness flowed through human hands and minds. The human capacity for curiosity and invention, combined with the ability to create symbols and assign meaning, led a reclusive Himalayan waif to every continent on the planet and into the moral consciousness and expanding lawns of urbanizing society. No other plant species has realized this level of ecological success. That such an otherwise unpromising, sexually repressed species would ultimately be so admired and despised is a testament to the unwitting contribution of human creativity and contradictions. One might think that in the view of Dandelion, it couldn't get any better than this. But there's a chance that Dandelion, with the aid of its coevolutionary ally, might get a further genetic boost that could make it more robust, more fit, and more successful.

Bouncing Forward and Back

Sitting in my colleague Matt's office on the campus of the Ohio Agricultural Research and Development Center, I looked at a zip-loc bag dangled in front of me. It bulged with some dried cottony plant material. "Ever hear of cis-1,4-polyisoprene?" Matt asked. In the verbal sparring between us I rarely scored a point, especially when it came to fancy chemistry, so I just smiled and listened.

"It's a critical plant-based resource that makes modern life possible. It's in thousands of consumer goods. Synthetic versions aren't good enough for products that require 100% natural, plant-based cis-1,4-polyisoprene. Current production comes from Malaysia and Indonesia, where unstable governments could cut off the supply and cause economic chaos." All this delivered without a pause. "These seeds"—he pointed to the bag—"are from a plant that produces high quality cis-1,4-polyisoprene. An industry consortium wants us to figure out how to grow it in Ohio."

I couldn't hold out much longer, so I showed my hand: "Does it have the quality necessary for condoms?" Matt was quick: "Yes. And the tires on Air Force jets are also 100% natural rubber.[38] If we don't develop this crop, there will be a population explosion and no way to protect it." I had one more question: "How does this involve me?"

The bag contained seeds of *Taraxacum kok-saghyz*, a relative of weedy dandelion. This *Taraxacum*, called "TK," produces rubber in its roots. It comes from the Tien Shan mountain range that runs from China into Kazakhstan. The Soviets thought it might be useful and cultivated it in the 1930s. It was tested in the United States in the 1940s, amid concern about wartime shortages of natural rubber.[39] Cultivation was abandoned after the war, when access was restored to Para rubber trees (*Hevea brasiliensis*) in Asia. The world demand for high-performance tires, gloves, medical devices, and thousands of other products, depends on Para rubber. That demand is increasing; shortfalls are predicted within the decade. Commercial and scientific interest in finding an alternative source of natural rubber has led to TK, dandelion's cousin.[40]

Seeds in the bag came to involve me because I'm the weeds guy. I might know something about the biology of a plant related to a weed. I wasn't sure. I couldn't assume that TK would behave like the dandelion I know. Many biological and technical obstacles must be overcome to domesticate TK as an economically viable crop. But the biggest obstacle arises every time we talk about it to audiences of potential investors, farmers, suburbanites, and the rubber-consuming public. Immediately, hands go up. The questions are not about the plant's adaptation to our soils or its water requirements. Instead: "Is this going to become another weed in my lawn?"

The answer is almost certainly "No." Almost. The seeds in the zip-loc bag will not become a weed, because TK doesn't behave as a weed. It germinates and grows slowly. It puts resources into making rubber, not leafy biomass. A better question is whether TK and common dandelion can cross-pollinate to make a new hybrid weed. If so, the hybrid seeds would have unique combinations of genetic traits. They might have the weed's genes for fast growth and large roots plus TK's genes for drought resistance and insect tolerance. Hybridization might make a weedy plant that does not yet exist on the landscape.[41]

Lucky for us, weedy dandelion is apomictic. Their interrupted meiosis squelches genetic recombination. They won't be fertilized by pollen from TK.[42] But flower sex goes both ways. Can pollen from weedy dandelion be accepted by TK plants to make hybrid dandelions that will take over the world? It's unlikely, due to basic differences in fitness and breeding systems of the two species.[43] Moreover, selective breeding is making TK more croplike, and less weedlike. TK will behave like a crop—germinating, growing, flowering, and maturing uniformly and predictably. In the meantime, research groups in Europe, China, and North America are planting fields of genetically diverse TK, hoping to select adapted, vigorous types with high rubber production. Pollen is moving in the wind. Seeds are floating in the air.

When those hands go up with the inevitable question, everybody looks over at me, the weeds guy. I could launch into a lecture about differential breeding systems that block hybridization. I could explicate environmental reasons why another species in the lawn is not a bad thing. Instead, I restate the obvious: the idea that the two species might share the landscape, the possibility that they would pump out thousands of shiny new genetically novel hybrid dandelion plants, the dread that hybrids would be even more troublesome (or glorious depending on your perspective) than those already gracing lawns across America, the fear that their seeds would float all around—well, basically, it scares the hell out of people. Dandelion, after all, is not just a plant. The acceptability of cousin TK as a crop will be determined not by complicated biology, or what breeders do to it, but by how it plays on the stage of the human imagination.

I got up to leave Matt's office. He finally gave in: "By the way, how did you know what cis-polyisoprene is used for?" Not, of course, from chemistry class. I learned about it from my father back in high school. We lived outside of Akron, then the rubber capital of the world, and he worked for a tire company. I went with siblings and neighborhood kids down to the end of the street to wait for the school bus on still mornings. The sky in the distance darkened with stinky, greasy, soot of long polymers spewing from the rubber factories. We got a daily dose of smelly hydrocarbons in the morning air courtesy of the rubber industry, and I got chemistry lectures from my father. So cis-polyisoprene is in my DNA, maybe in more ways than I want to know.

Florida beggarweed, *Desmodium tortuosum*.

2

FLORIDA BEGGARWEED

The scene outside Makola Market in Ghana's capital, Accra, pulsed to the rhythm of West African highlife. Streets bustled with grey and yellow taxis, rickety bicycles, horn-blowing minivans, load-weary trucks, and wooden wagons pulled by sweat-dripping men in dark shorts and flip-flops. Women in long wraps and tall Hausa men in striped frocks pushed along the narrow sidewalk while children darted in and out balancing platters of bread, dried fish, or bananas. A throbbing sun pulled smells of smoke, sweat, and spoilage into the hazy ocean air.

I visited Makola every week during my Peace Corps volunteer stint in the 1970s. With rudimentary Twi I bargained for things like okra, plantain, and cocoyam leaves—staples of my Americanized version of West African cooking. One day near the end of my tour, I had another objective. After nearly two years away from home, I craved real American food. A market woman just returned from the United States told me of a certain seller at the far end of the line of kiosks, who had what my taste buds yearned for.

I squeezed past rows of stalls. Tin roofs cast patches of shade. The sellers—all women—were large, loud, and brightly clothed in blazing prints with matching scarves. Banter was constant, syllables clipped, the air punctuated with pointing, snapping, slashing fingers. Around the corner a woman sweltered beside an immense kettle over charcoal flames. In her hands, a long wooden paddle roiled a sea of roasting shelled peanuts. The meaty-popcorn-spicy-burnt-caramel-like smell took me back to the Nut Shop on Main Street in my Ohio town at Christmastime. Two young women in matching yellow wraps, one with a baby on her back, pounded long tree-branch pestles in a wooden mortar. Next to them, the promised seller: on the shy side of twelve years, hair neatly plaited above shining almond eyes, and barely visible behind a large, lumpy mound of gloriously sticky peanut butter. For sixty Pesewas she waved at the flies, stabbed the mound with a spatula, and filled a plastic container. "Please, sir," she smiled, "Thank you, sir."

I lived and worked about twenty kilometers (twelve miles) outside of Accra, in the quiet village of Achimota. My job was vaguely described as an agronomist at Ghana's Animal Research Institute. I worked closely with Kobe, a thoughtful, square-jawed pasture expert from the Ashanti Region, who conducted field studies on forage crops—plants that cows and sheep eat either fresh in pastures or dried as hay. When not lampooning each other's culture, food, and manner of speaking, Kobe and I evaluated forage plants that might survive the hot, dry climate. We were especially interested in finding legumes that would provide good pasture and help to improve the soil.

I spent hours in the institute's library reading up on tropical legumes. I assembled a list of species for Kobe to consider. He caught me over lunch one day behind a textbook about forage crops. I showed him the chapter on plants for hot climates. We had already tested some of those species, but others were unfamiliar. One genus stood out to me as especially hopeful: *Desmodium*. It had large, attractive leaves, erect or trailing stems, and several of them were already being used in the tropics. Kobe winced and shook his head: "It makes me itch just to think of it." No further explanation. Instead, he pointed at my lunch and joked about the American practice of working while eating. It was the first peanut butter sandwich I'd had in two years.

Stepping into the Fog

The next time *Desmodium*, peanuts, and I were together was about ten years later, on the sprawling coastal plain of South Georgia. It was my first day as a research agronomist with USDA. I was in a pickup with CW, a good-natured agricultural extension (outreach education) expert of many years. He tutored me in peanut agriculture. Hardly mentioning the crop, he pointed to patches of Florida beggarweed, *Desmodium tortuosum*, in just about every field we passed. It was one of the species Kobe had rejected. To CW, Florida beggarweed was the most troublesome weed in southern peanut fields. Beggarweed infested corn, cotton, and soybean fields, too. But peanuts were a finicky crop and that's where it was worst. Farmers spent a lot of time and money trying to control beggarweed by cultivating, mowing, and spraying—lots of spraying.

We turned off the main road past a gas station intersection known as the town of Sycamore. The truck bumped down a farm lane, and stopped at a field of bushy, blotchy beggarweed. I stepped out toward the tall, brown-purple stems that seemed to inch higher before me. Floppy leaves drooped in the stifling heat, and branches lurched in all directions from the base of the plants. Small pink-purple flowers were spaced along ends of wiry branch tips. Seed pods were thin, twisted, and divided into three to five segments, each holding a single seed. The whole plant was covered in short, sticky, itchy hairs that oozed a thin oil and diffracted sunlight into a grey fog. Beggarweeds towered over the peanut plants, shading them with a rough jumble of stems, leaves, and seed pods that reached out to grasp passing animals, tangle in machinery, and ensnare pant legs of anyone unlucky enough to walk through a patch of them, or have the misfortune to begin their research career working with them. Kobe had warned me about this.

A spray rig crawled across the next field, a rumbling diesel engine erupted down the lane, and an irrigation system came to life in the distance. The crop, the soil, the weeds, the gnats pestering my face and neck, the smell of insecticides from a nearby cotton field—it all seemed a little unsettling. I didn't tell CW this was my first time in a peanut field, or that as an Ohio kid fresh out of a northern graduate school I wasn't sure I was the right person to take on the tough weeds in peanuts and cotton. Likely, the thought crossed his mind.

I had done my homework before moving to Georgia. I read up on peanuts and Florida beggarweed. I even found a copy of the textbook Kobe and I had looked at a decade back. Now here it was in front of me. Of the three thousand or so plant species that made a home on this landscape, agricultural practices had whittled them down to just two that vied for the same sunlight—peanuts, planted and nurtured, and beggarweed, intruding and thriving. It hadn't always been this way, CW explained. Beggarweed had been in the background for years. Nobody thought it would ever be a problem. But now it was the weed that farmers grumbled about the most.

A history professor had advised that some kinds of problems could be fixed by understanding how they came to be. A proud history was part of the verbal landscape in South Georgia. But that didn't seem to fix the situation with Florida beggarweed. By the time the weed population was big enough to be a problem, "the encroachment couldn't be undone," as CW explained. However it got there, whatever allowed it to spread, whatever made it more of a problem today than in the past, whatever allowed it to hold on in these hot, droughty soils, in the face of all the efforts directed against it—well, the time to worry about all those things was long gone. "I guess this is your weed now," CW said with half a smile.

Links in the Chain

The Georgia Coastal Plain Experiment Station opened in 1919 as the first agriculture research site on the deep sandy soils that stretched from Delaware to the Gulf Coast. The station was set on a gently sloping plain of dry scrubby brush that, when cleared, leveled, and irrigated, gave rise to patches of cotton, peanuts, corn, soybeans, tobacco, vegetables, and groves of sprawling pecan trees. Fields blanketed by wheat, rye, and oats in winter reawakened with watermelons and vegetables by June. Long-armed irrigation systems swept the ground, sending up sprays that caught the sun in a rainbow of mist. Crop dusting planes skittered across fields, swooping up sharply before telephone wires or stands of loblolly pine. In late summer, cotton fields turned to white while peanut plants tanned in the sun, their roots upturned in neat rows. On dry autumn days, mammoth combines growled across the earth ahead of rolling clouds of red dust and dirt.

In the center of this landscape rested the town of Tifton, a motel-stop along US 75, the western route to winter homes in Florida. Locals assured me that Tifton was a great place to live, a great place to raise a family. They were proud of their slow pace and neighborliness. You could still get a twist cone for fifty cents at the Sugar Shack on Love Avenue, down from the cinema. Folks enjoyed friendly ribbing, boastful one-liners, and whoppers that rolled out thick as hair on a dog's back. And by the way, what church do you go to?

I knew it would take a while to appreciate the local ways. Until then things might be unclear. I got the feeling there was something beneath the surface, something the locals knew that I might or might not understand over time. People seemed self-assured, but I wondered if I made them uneasy with my Yankee speech pattern and general cluelessness. Folks talked a lot about pride, being Southern, their proud Southern heritage.

Beggarweed was just as hard to understand. I pored over the plant taxonomy books and found that its origin and classification were confusing even to experts. They disagreed over where the plant came from and what it should be called. The common name was not peanut-weed or drought-weed; it was a name connected to place and people: Florida beggarweed. A hot place and desperate people. Locals often called it "beggar-lice," no more promising. To the Portuguese it was "pega-pega," catch-and-release, a story there to be sure. And the scientific name, *Desmodium tortuosum*, wasn't any more helpful. The word *Desmodium* was from the Greek word for "chain," referring possibly to the segments of the seed pods, or maybe the way the plants clung together.[1] A chain and *tortuosum*—the name signaled something twisted, convoluted, or worse. I tried not to think too much about what I had gotten myself into.

I soon found that the best place to learn about beggarweed and local agriculture was the Southern Oven Diner. Farmers lingered over biscuits and gravy, hot grits and butter, and sweet or unsweet tea, to chortle and grumble about business. They complained of the big loans needed to get a crop in the ground. They grimaced as hoped-for profits dwindled with drought, hailstorms, white mold, and leaf blight. They anguished over weeds that popped up in peanut fields. If everything went right, the weeds would be held back so the crop could grow and secure their investments in seeds, fertilizers, irrigation, pesticides, land rental, labor, and insurance. Often, things didn't go right, and weeds evaded their control program. Beggarweed was the worst of the bunch because it grew fast and tall, sneaking like a snake through openings in the crop canopy, stealing light, water, and nutrients.

Farmers' biggest complaint was a lack of options. "Not enough tools in the toolbox," a phrase I heard over and over. The one tool they all seemed to use was a chemical weed killer called dinoseb herbicide. They'd used dinoseb for twenty-five years to kill the weeds infesting peanut fields.[2] They sprayed dinoseb when peanut seedlings first cracked through the soil surface. If conditions were good, small weeds died and peanuts survived. Dinoseb was the main herbicide used to kill beggarweed and other broadleaf weeds. It might kill 75 percent of the beggarweed plants, and on a good day 85 percent. That meant the other 15 to 25 percent—the tougher ones—had more room to grow. Some were inherently more tolerant than others. The stronger Florida beggarweed plants were among the last weeds standing, flowering, and spreading seeds. Farmers wanted more options, more tools to deal with beggarweed. By this, I knew, they meant more, better, herbicides.

Spray-and-pray is what the professionals called it. Most people doing research on weeds at that time sprayed a lot of herbicides in test plots and prayed that one of them killed the target weed without harming the crop. Some of the faithful enjoyed the sense of control over the plant world. I was lucky that my bosses at USDA encouraged me to explore other approaches to manage weeds in peanut fields. Colleagues who studied insects and plant pathogens were developing ways to reduce the overuse of pesticides. I wanted to do the same for weeds. That meant I had to know more about Florida beggarweed than just how to kill it with a cocktail of chemicals.

I drove out to the research farm every few days to record the progress of the struggle between beggarweed and peanuts. Though they are botanical cousins, the two species are not natural allies. They are competitors for the same meager resources of the droughty coastal plain soils. Both belong to the legume family, a large group of plants that's been around for about 60 million years.[3] And both species capture, or "fix," needed nitrogen with the help of some soil bacteria and thus thrive in fields having soils of low fertility. Deep sand, low nutrient availability, frequent tillage, annual harvest and clearing, high temperatures, high humidity, unrelenting sun— this was the perfect environment for legumes of tropical origin. There are thousands of tropical legume species, including several hundred plants in the *Desmodium* genus. Somehow, only this one, beggarweed, had found its way to fields where farmers planted peanuts.

Beggarweed and peanuts were like gears churning in a vast agricultural machine. Farmers wanted more tools to keep the machine running. I knew the aquifer that irrigated those fields was not unlimited. Consumers wanted food and fiber without those chemical tools. Fertilizers and pesticides moved with soil erosion. But it wasn't part of the farm (or agricultural research) culture to question how long the machine could keep running. I needed to maintain good connections with the peanut industry. But I wanted my research to focus on weeds in a future where social, economic, and environmental pressures would make current tools obsolete. I knew the coastal plain station was a great place to work with weeds, crops, and farmers. I also knew that working as an outsider to change a flawed machine would put me in league with sinners who roll boulders uphill.

Trichomes Can't Be Choosers

Florida beggarweed and peanuts came from South America, but it's likely they never met each other there. They are from very different, isolated eco-zones separated by the Amazon basin. Peanuts, *Arachis hypogea*, are rooted in the humid southwestern forests of Brazil and Paraguay.[4] Florida beggarweed arose from the hot, dry, sparse northeastern coastal desert.[5]

The shape and size of the two species are about as different as any two leguminous plants. Peanut plants are low-lying perennials whose stems creep along the ground. They put their flowers on branches that hug the soil and send a stemlike peg into the dirt to produce a swollen pod with seeds, or nuts, inside. In contrast, beggarweed features tall lurching branches that hold delicate pea-shaped flowers along upper stems. Instead of hiding seeds underground and waiting for humans to uncover them, Florida beggarweed seed pods hang along the ends of limp stems. The seed pods are divided into one-seeded segments that separate from one another.

It's like snapping a string bean into many pieces, each with a seed inside. Beggarweed seed pods are covered with thousands of short, hooked hairs called trichomes. The trichomes secrete gluey resins that help the seed pod segments attach to skin, hair, and clothing. Tacky seed pods dangle from loose, wiry branches that wave in the wind, begging to be carried away to unexpected parts of the world.

Peanut plants caught the interest of humans around four thousand years ago.[6] Woodland foragers found and collected them while digging for edible tubers. Eaten fresh, dried, or after being thrown into hot coals, peanuts became the ultimate low-tech snack food, easy to carry in a biodegradable package. Local people dispersed them across South America to the Caribbean and Mexico. In some places they were exalted, buried with the dead, used in sacrificial offerings, and tossed into fires or rivers to be carried to the gods.[7]

Beggarweed became an annoyance when animal herders roamed the dry, sparse grazing land. Beggarweed seeds, in their sticky pods, got caught in hair or intestinal tracts. The seeds are small and unpalatable, but grazing animals found the fleshy leaves and young stems to be a rare source of nutrition.

Over generations, hunter-gatherers selected peanut genotypes with many large seeds that tasted good. This led to domestication of the crop. Meanwhile, on the other side of the continent, herders and their stock dispersed seeds of wild beggarweed. Unconsciously selected genotypes were those with especially adherent seeds. In this way, beggarweed was partly domesticated. It hooked itself to human activity without losing its adaptive wild traits: erratic germination, long seed dormancy, uneven flowering, and unrestrained dispersal.

People and plants of beggarweed's arid northeast homeland were physically and culturally disconnected from those of peanut's moist southwest. The two species remained contentedly isolated from each other and the rest of the world for hundreds of years.

Buried Nut and Buried History

I got to know the coastal plain by driving the back roads through small towns. I was intrigued by those that claimed to be the Peanut Capital of the state, the country, the world. Most held a peanut festival with a parade and crowning of a blonde, adolescent Peanut Queen. Peanut statues adorned town squares. In Early County, Georgia, for example, a marble peanut monument across the green from the Confederate flagpole was engraved with a humble tribute to the humble plant. None of the sculptures, plaques, or free literature said much about the crop's heritage. No statues honored heroic figures who brought peanuts to American shores. But it was clear that peanuts had a proud history steeped in American values of liberty and freedom. I kept hoping for an exhibit or reenactment showing how the crop journeyed from the center of South America to the North American coast.

A historical memorial to peanuts, as well as beggarweed and all South American crops, weeds, and people would start with the Columbian Exchange. Peanuts and beggarweed could not escape the political, cultural, and ecological chaos that sent European explorers to new lands rich with stores of gold, timber, diverse crops, and people.

Portuguese explorers aiming for East Asia landed in what became Brazil in early 1500s. They encountered peanuts growing in the region where the crop was first domesticated. Portuguese naturalist Gabriel Soares de Souza described early peanut cultivation: "The husbands know nothing about these labors, if the husbands or their male slaves were to plant them, they would not sprout."[8] Eventually, Portuguese sailors working trade routes to Brazil and the West Indies carried peanuts to Africa and Western Europe.[9]

European sailors plying African coastlines came to rely on local African crops like red rice, millet, sorghum, sesame, palm oil, and others. In return, they delivered to Africa a cornucopia of crop plants they had collected in South America, including maize, cassava, squash, beans, tomatoes, potatoes, and tobacco.[10] This is how peanuts arrived in Africa along the western "Upper Guinea" coast.[11] The time is uncertain.[12] Peanut cultivation spread quickly into the interior and peanuts moved with trade in gold, ivory, spices, and enslaved people across the Sahara and into southern and eastern forests. Peanuts became so widespread in Africa that botanists long believed they originated on that continent.[13]

Expansive farms were needed to feed all the people engaged in the complex system that delivered enslaved Africans to the Atlantic trade system.[14] The European overseers had no experience with tropical crops and soils, which are tricky to manage. So they relied on the expertise of the Africans, whose knowledge had been passed down through generations of farmers.[15]

   This pattern was repeated when enslaved African agriculturalists were transported to the West Indies and South America. Enslaved Africans already knew how to grow the indigenous crops like cassava, maize, and peanuts. African farmers adapted their agricultural knowledge to the New World tropics. Neither they nor their captors would have survived without their expertise.[16]

   In exchange for stolen treasure, Europeans delivered smallpox, measles, and cholera to the New World.[17] Indigenous populations were quickly decimated. This left enslaved African agriculturalists, tolerant of these diseases, to maintain the agricultural systems, including crops, soils, and their management. Enslaved Africans became the guardians of the diverse crop germplasm and stewards of cropping practices in the American tropics.[18] Among the crops were peanuts. By the mid-1600s, peanuts had found their way to Spain, Portugal, the Philippines, and China. They had not yet reached North America.[19]

Love-Herb Meets a Fistful of Hope

People engaged in the triangular trade system had little reason to give beggarweed any attention except as an annoyance. Its journey to southeastern North America is obscure. The earliest descriptions I've been able to find came from the Irish born naturalist and physician, Hans Sloane. In 1687, Sloane sailed to the West Indies hoping to discover new drugs. He documented many animals and objects used by indigenous peoples and Africans, including nearly eight hundred plant species new to Europeans. He described them in a two-volume work, published in 1707 and 1725, with the catchy title, *A Voyage to the Islands Madera, Barbados, Nieves, S. Christophers, and Jamaica, with the Natural History of the Herbs and Trees, Four-footed Beasts, Fishes, Birds, Insects, Reptiles, etc. of the last of those Islands, to which is prefix'd an Introduction Wherein is an Account of the Inhabitants, Air, Waters, Diseases, Trade and etc. of that Place, with some Relations concerning the Neighboring Continent, and Islands of America.*[20] The illustrations are as luxurious as the title.

The first recorded intersection of peanuts, beggarweed, and slavery is tucked in the pages of Sloane's writings. His descriptions of peanuts and beggarweed on Jamaica showed that the two species shared the same island if not necessarily the same habitat. He explained the essential role that peanuts played in the transport of enslaved Africans: "The fruit, which are call'd by Seamen Earth-Nuts, are brought from Guinea in the Negroes Ships, to feed the Negroes withal in their Voyage from Guinea to Jamaica." Sloane explained that the nut was used by the Portuguese to "victual their slaves to be carried from St. Thomas to Lisbon." This confirms movement of the plant and the people in both directions across the Atlantic.[21]

Sloane was first to link beggarweed to agricultural crops, not just grazing land. He described three plant specimens using the old name, *Hedysarum triphyllum,* variants of what became known as *Desmodium tortuosum.*[22] One of his drawings (plate 116) is unmistakably Florida beggarweed. Sloane found it growing near the Town of St. Jago de la Vega, "and in the paths among the sugarcanes in several places of this Island." Sloane, a physician, found no medicinal value in the plant. But he made note of the segmented fruits, "each joint of which is fastened to that next it, by a very small isthmus, whereby its adhesion to it is so easy, that by its roughness sticking to any garment, they leave one another, whence the Portuguese name *Erva d'Amor*" (love-herb).[23] This is the earliest description of beggarweed in a crop field, and of the mechanism for its dispersal between Africa and the West Indies. Thus, peanuts and the love-herb both crossed the ocean on slave ships, the first as essential sustenance for the journey, the second sticking to garments, skin, or hair as an unnoticed castaway.

Beggarweed, like peanuts, became so widespread in Africa that European botanists thought it originated there. Biogeographer Charles Pickering, writing in the mid-1800s, put its origin in Equatorial Africa and connected it with European colonization by way of the West Indies. He described "*Desmodium tortuosum* . . . a shrubby Leguminous plant through European colonists carried to the West Indies and observed by Sloane . . . on Jamaica, often in cultivated ground . . . Eastward, is known to grow . . . from Senegal to Abyssinia" (Ethiopia).[24] Botanist William Jackson Hooker mentioned beggarweed in a tome with a title that puts Sloane to shame: *Niger flora; or, An enumeration of the plants of western tropical Africa, collected by the late Dr. Theodore Vogel, botanist to the voyage of the expedition sent by Her Britannic Majesty to the river Niger in 1841, under the command of Capt. H.D. Trotter, R.N., &c.; including Spicilegia gorgonea, by P.B. Webb, esq., and Flora nigritiana, by Dr. J.D. Hooker . . . and George Bentham, esq. with a sketch of the life of Dr. Vogel. Ed. by Sir W.J. Hooker . . . With two views, a map, and fifty plates.*[25] The book never made the bestsellers list, but it suggested that beggarweed was widespread, variable, and had been in West Africa a long time.

To early European botanists, Florida beggarweed was just one of hundreds of species with little value as food, medicine, or ornament. Pickering took note of it in the Americas and put its origin in Africa. Vogel recorded it in West Africa, but Hooker described it as common in tropical America. Alphonse de Candolle said the origin was uncertain, "shared between Africa and America," in his 1855 work, briefly titled *Géographie botanique raisonnée; ou, Exposition des faits principaux et des lois concernant la distribution géographique des plantes de l'époque actuelle.*[26] The Swedish botanist Olof Swartz, the authority for the scientific name of the species, did not even mention Africa in his description of the plant.[27] He had spent his exploring years in the West Indies and never visited Africa. After all, these early botanists could not go everywhere, and their time had to have been occupied doing other tedious things, like reading the titles of each others' books.

How, when, and by whom peanuts and beggarweed got to North America is unknown. There's little doubt they arrived with the growing traffic of ships moving between Europe, Africa, and North America carrying human cargo to support cotton, tobacco, and sugarcane production in the colonies. In an otherwise comprehensive book about peanuts, we read that "eventually it [the peanut] traveled to the colonial seaboard of the present southeastern United States, but the time and place of introduction was not documented."

In a world of chaos, lives torn apart, surrounded by brutality, what hope remained was held in a fist full of seeds that might ensure survival, maintain a culture, and become a major food crop. There are stories that the seeds were hidden away in hair or clothing.[28] Nothing is known of who among the cargo had the courage and wisdom to bring along the seeds of home. I used to gaze across the vast patchwork of weedy peanut fields when I worked in the hot southern coastal plain. Sometimes, sitting in my air-conditioned car, I struggled to connect with the landscape, knowing that seeds of the plants before me had been carried by half-starving, half-naked, frightened human beings whose bodies—but not spirit—had been stolen and chained to the hull of a ship.

For beggarweed, the journey was less poignant but equally ambiguous. Useless as human food, it was most likely thrown onto ships for animal fodder. With its grasping trichomes, there would have been no need for anyone to sneak the seeds onboard, even if it had some use or significance. Peanuts and beggarweed were not alone in their association with human bondage: almost every major crop and weed has such a connection. Southern states have slowly abandoned flags that glorify cultures of slavery. It would be impossible to abandon the foods and fibers that serve as reminders and continue to demand constant weeding.[29]

Changing Fortunes over America's First Century

For one hundred years after arriving in America, peanuts remained a crop of mostly enslaved Africans. A 1769 report indicated that peanuts had been "well known" in the colonies of the South, where they were planted "in small patches by Negroes for market." Peanuts were sold and traded among those of African heritage, ground into a meal or eaten whole after boiling or roasting. Enslaved servants of French Creole refugees from the Haitian insurrection brought peanuts with them when they escaped to Philadelphia in 1791. They prepared them in cakes and soup.[30]

Peanuts became known outside of the African heritage community along the mid-Atlantic coast. In the 1780s, Thomas Jefferson, defender of human liberty and bondage, planted "peendars" and found them to be "sweet."[31] Most plantation owners, however, scoffed at the "slave food." Some grew peanuts for hog feed to produce peanut-flavored pork, a delicacy. Others grew them as a green manure crop, turning them into the soil to return nutrients before hogs were let in to root out the remaining nuts.[32]

There are no records of beggarweed in North American agriculture during colonial times. The American naturalist, William Bartram, walking along the Suwanee River from Georgia into Florida in the late 1770s, recorded a plant of the old genus name *Hedysarum*. If this was the plant that became Florida beggarweed, Bartram's mention of it was the earliest record in North America.[33] Neither he, his botanist father, nor accounts of early settlers mentioned it. The annoying stick-tight seeds would certainly have caught their attention. The oldest specimen in a US herbarium is a relatively late (1843) collection from Leon County, Florida (around Tallahassee), although it had been a fixture of the Deep South by that time.[34]

While beggarweed languished in the Florida heat, peanuts were confined to small plots up around the Mason-Dixon Line. The land between them was a mix of pine trees, scraggly shrubs, and dank swamps. It was also home to the Creek and Cherokee people. What set the stage for peanuts and beggarweed to come together in North America began around 1813 with Andrew Jackson's wars against people of the Creek Nation. Species-diverse forests were destroyed, villages burned, thousands of men, women, and children killed or displaced.[35] From the Carolinas to Florida and west to Texas, wealthy would-be plantation owners (and Jackson relatives) bought up the tillable real estate in Jackson-land for tobacco, cotton, and human subjugation.

In his second message to Congress (1830) Jackson explained that "toward the aborigines of the country no one can indulge a more friendly feeling than myself."[36] Later, Martin Van Buren, who oversaw the Trail of Tears, congratulated himself in his 1838 report to Congress: "The entire removal of the Cherokee Nation to their new homes west of the Mississippi . . . have had the happiest effects."[37] For the future of peanuts and Florida beggarweed, nothing could have had a happier effect; their paths to future ecological success were assured.

Horse- and mule-drawn and human-exploited agriculture quickly transformed the landscape and culture of the South.[38] But never mind the humans: those horses and mules needed something to eat. The local grasses could provide energy for working animals, but not the protein they needed. In the humid south, protein-rich temperate forages like alfalfa and red clover do not survive. What farmers needed was a productive, high-quality, nitrogen-fixing forage legume that could withstand the heat and drought.

In St. Augustine, Florida, a farmer named John Latham offered a solution. Latham wrote a letter to the *Southern Agriculturist*: "I send for your inspection a species . . . that horses and cows both very greedily

devour. . . . It grows wild in several parts of Florida, and in the old fields."
The editor identified the plant as *Hedysarum,* adding, "We are not acquainted with its properties as fodder for cattle," and advised Latham to
plant one of the native grasses instead.[39] We don't know if Latham took
the editor's advice. Farmers generally did not. Florida beggarweed soon became a valued source of feed for the horses that pulled plows through soils
across the Deep South with the hand of forced labor on the stilt.

Latham's letter revealed more of beggarweed's nature. "It is called
by negroes Sweet Weed, why, I know not," he wrote. Thus, enslaved
Africans apparently knew it well. In spite of reports to the contrary,
it might have been useful pharmacologically. A recent review of *Des-
modiums* indicates that people in Africa and Latin America have used
Florida beggarweed to treat uterine fibroids; leaves and stems were
ground to make a drink, "to cure menstruation pain, stomach pain, and
for psychomotor development."[40] Moreover, a study of traditional African drugs indicated that Florida beggarweed was used to "attain good
luck."[41] The meaning of this was not explained in the study. The connection of "sweet weed" with curing menstrual pain and attaining luck
hints of deeper knowledge of Florida beggarweed than was recorded by
European botanists.

By the mid-1800s, beggarweed was on its way to becoming the premiere forage crop of the South. Without commercial production or
sales outlets, the seeds were passed on informally among farmers. The
plant spread slowly through counties along the coastal plain from the
Florida panhandle west toward Texas and up through Georgia toward
the Carolinas. Beggarweed provided fuel for beasts of burden and so-
lidified its reputation as a valuable hay crop that horses, especially,
loved to eat.

Meanwhile, peanuts were on the way to tepid acceptability. An 1860
article in the *Southern Cultivator,* which promoted the South's agricultural independence, gave instructions for cultivation of the "ground pea,
pinder, goober peas etc." for human consumption rather than as fodder
or hog feed. The *Cultivator* was the most influential southern agricultural
paper from the 1840s to 1860s. Since it had a mostly white readership,
the article is an early suggestion of a change in attitude toward the peanut.

With beggarweed powering animal-drawn machines, the whole country became wrapped up in the cotton economy. New York City became

the hub of cotton shipping and merchandising. Some called it the capital of the South, reaping about 40 percent of cotton profits estimated around $200 million annually.[42] When Lincoln was elected, New York Mayor Fernando Wood asserted that if the Union dissolved, the city should declare itself independent and continue trade with the South.[43]

Civil War finally changed the status of the peanut. Nuts were carried and consumed by soldiers on both sides. The song, "Eatin' Goober Peas" suggested they were a mainstay among Confederate soldiers. When Union forces cut supplies to the Confederate army at Petersburg and Richmond in 1865, General Robert E. Lee complained that his appeals for help from the Confederate Congress were ignored: "They do not seem able to do anything except eat peanuts and chew tobacco, while my army is starving."[44]

Peanuts became more widely known when freedmen sold them raw, boiled, and roasted in markets in the South and North. They moved from gardens to production fields when the profitability of cotton declined in the mid- to late-1800s. Cotton cultivation had degraded the soil, labor productivity had fallen, and Britain was buying cheaper cotton from India.[45] The war had devastated Southern agriculture, and recovery was slow. The *Southern Cultivator* advocated peanuts as a crop for "cotton-poor" farmers. In 1868, Nicholas Nixon of Porter's Neck Plantation near Wilmington, North Carolina, published a long article describing methods for growing peanuts on a large scale. Nixon had been developing peanut production methods for several decades using knowledge gained from previously enslaved Africans.[46]

Media representation of peanuts was slow to change. An 1870 article in *Harper's Weekly* opined that polite society still considered the eating of peanuts to be "ungenteel." Peanuts were associated with rowdy behavior at sporting events and theaters where the shells were a particular nuisance.[47] In response, folks in Wilmington, North Carolina, dispatched street vendors to hawk hot, roasted peanuts to reach a wider—and whiter—range of consumers. Twenty years later, Rettia Anne Spence blessed peanut consumption by genteel folks in a *Good Housekeeping* article on the "Social Rise of the Peanut." Along with stories of "College Girl Manners" and "Starching and Ironing," she reassured readers that the peanut "travels on its merits and comes in contact with the best people of our land."[48]

Southern Salvation and Glory

The fortunes of Florida beggarweed rose quickly as it aided the South's recovery. The *Georgia Weekly Telegraph* said beggarweed was "unknown in South Georgia and North Florida until the war, when it began to be observed in fields where herds of South Florida cattle had been pastured." It was "cultivated by many in the corn-fields" where it "grows from two to six feet high, and, on rich land, forms a dense vegetation, which almost hides the corn from sight." Hogs, cattle, and horses ate it "voraciously, even neglecting the standing corn" to consume it. In addition, it was grown "as a fertilizer for worn lands . . . and has the advantage . . . that . . . it seeds itself from year to year." Bales of it were to be exhibited at the upcoming State Fair.[49]

The hopeful, bushy species claimed public favor. The *Southern Cultivator* published a letter to the editor entitled, "Indian clover or beggar weed" written by "S. W. B." from Brooks County, Georgia.[50] He saluted the plant for "its great value as a forage crop, for its immense yield per acre, and for its nutritious qualities for all kinds of stock." It produced the "most delicious hay, thousands of pounds per acre," a yield so abundant that "there never need to be another poor milk cow from the seaboard to the blue-ridge." Readers could view samples at the Agricultural Meeting at Gainesville (forerunner to the state fair), the Agricultural Department in Atlanta, or Colonel J. H. Martin in Columbus, Georgia. A letter in the next issue of the *Cultivator* by Geo. M. Bates of Lake City, Florida, advised farmers to "have it seeded broadcast thickly" for feed as well as "a good fertilizer."

Beggarweed helped attract northerners to Florida. In 1888, The *Daily Inter Ocean* of Chicago published the story of Captain John W. McRae, who had gone to the Sunshine State in poor health, with "less than $1000 . . . no horse for the first two years [and] was expected by his relatives to die of consumption." But Captain McRae "now owns one of the best stores in the town, numerous lots of land . . ." on which he cultivates "wonderful beggar-weed, which grows wild . . . and is claimed to be the finest fodder plant feed in the world for horses."[51]

Within ten years, news of this valuable plant had spread toward the northern limit of its adaptation. Newspaper reports of the 1897 North

Carolina State Fair called attention to "one of the most interesting exhibits," that of Sergeant Hamilton who "lost his eye sight in the service of his country, but . . . his pluck and patriotism have made him a successful farmer as well as a patriot of the first order." His collection included "specimens of the giant beggar weed, the new forage plant, nine feet high, [which] is said to improve the land faster than peas and clover [and has] more properties for fattening cattle than any other plant known." Also in his collection were "Spanish peanuts, noted as good for fattening hogs . . . and a fine exhibit of native wines."[52]

Florida beggarweed "never requires reseeding" according to the 1900 catalogue of the N. L. Willet Drug Company. Among ads for Dr. Gilder's liver pills, the issue described "Giant beggar weed." The crop "has been to Florida what clover has been to Tennessee and Kentucky, and Peas to Georgia and Alabama. But it is superior to either in that it will thrive on much poorer land . . . is more fattening to stock of all sorts . . . and enriches the land." Best of all, it "interferes with no crop [and is] easily kept under cultivation." Half a pound of seeds sold for 23 cents, about the price of a dozen eggs.

A statuesque woman in a flowing dress, one breast bare, blowing a long trumpet, was drawn on the cover of the catalogue of the Alexander Seed Company. She announced new varieties like "our improved 'Bon Air' rutabaga" along with "Giant Beggar Weed [which] grows in popularity each year. Sow at any time after frost is over, until the middle of June in drills . . . three to four pounds to the acre, or broadcast ten to twelve pounds to the acre. . . . [a] pound 45 cents postpaid."

For those who were not aroused by seed company encouragement, farmer testimonials, or drawings of bare-breasted women, the folks at the USDA Division of Agrostology issued Circular Number 13. Emphatically, Florida beggarweed "never becomes a weed."[53] With this assurance, the ecological success of Florida beggarweed was guaranteed. Seeds were spread throughout the Southeast and as far as it would grow to the north and west. It was promoted for its value as livestock feed and soil enhancer. The seeds continued to be advertised in catalogues into the 1960s. Praise for this plant appeared in Agriculture Experiment Station bulletins, publications, and textbooks about forage crops that made their way across the world.[54]

The Start of a Sticky Entanglement

Southern agriculture changed when a little bug appeared in Brownsville, Texas, in the 1890s.[55] Over the next twenty years, the Mexican boll weevil (*Anthonomus grandis*) ate its way north and east, reducing cotton yields, and bringing an end to the southern sharecropping system. Many farmers seeking alternative crops turned to peanuts just in time for the crop's new social acceptance.

Soon came peanut butter, smooth and crunchy. The North American product was introduced at the 1904 World's Fair in St. Louis.[56] The equivalent of peanut butter was known to the Incas three thousand years previously, and later to the Aztecs. The roasting and pounding I had witnessed in Makola market differed little from peanut butter preparation in West Africa in the sixteenth century, using methods borrowed from South America.[57] Nevertheless, a Canadian patent for a process to make peanut paste was issued in 1884 to Marcellus Edson, a pharmacist. The patent for a different process was issued in 1895 to Dr. John Harvey Kellogg, then superintendent of the Battle Creek Sanitarium, inventor of corn flakes, and crusader against carnal pleasures of all kinds. Neither patent mentioned the Incas or West Africans, making peanut butter the most quintessentially American of foods.

The popularity of peanuts led some processors to seek cheap imports. This threatened the domestic market, so growers appealed to Congress for help. Testifying before the Ways and Means Committee in 1921 was Dr. George Washington Carver, a former slave, whose many peanut inventions included flour, cake, shoe polish, pickles, rubber, coffee, cough syrup, face cream, and hundreds more.[58]

Farmers, Florida beggarweed, and peanuts ultimately became entangled in each other's fate when tractors chugged onto the farm in the 1920s. Horses and mules soon disappeared, and beggarweed was left standing on a landscape that no longer needed beggarweed forage. Fields once thick with beggarweed hay were plowed and planted to peanuts, cotton, corn, tobacco, and a new crop—soybeans. Traits that made Florida beggarweed a good forage crop—quick growth, high biomass, abundant seeds that survive for years in the soil—revealed a species that had no intention of leaving the land. The heyday of beggarweed as a hay crop was over. People who once cherished the sight of the bushy leaves and pulled the catching seed pod segments from their clothes with charity began to think that a plant with "weed" in its name might actually be—well—a weed.

Peanuts and Florida beggarweed joined in a struggle for resources on the low fertility, sandy soils. The legume cousins were quite different in origin, appearance, structure, and mechanisms for seed dispersal. But they were similar enough in physiology and environmental adaptation to compete for the same limited resources of light, water, nutrients, and human attention. As the lowly peanut gained economic importance, the gawky beggarweed became an object of scorn, a target of whatever devices and poisons farmers could throw at it.

Fortified by steel and petroleum, modern agriculture embraced technology as the source of progress. Fields of monoculture crops took the form of noisy, smoke-chugging machines. With mechanization, farmers hoped to control all the things that made crop production difficult—soil fertility, insects, diseases, weeds—from the seat of the tractor.

Florida beggarweed produces flowers and seeds through autonomous self-pollination. Thus, it had little genetic variation and limited potential for adaptation to its new status as an object of scorn. Its only hope to be more than an ordinary weed rested in its human accomplices, agriculture experts, and a little organic chemistry. With cheap labor gone, farmers tinkered with oils, solvents, acids, and other chemicals to do the nasty job of killing weeds. Most concoctions that killed weeds also killed or injured their crops. But at the end of World War II, the industry part of the military-industrial complex repurposed their expertise to develop chemicals for use in agriculture.[59]

Among the products were herbicides with astonishing powers. One of the powers was *selectivity*. Herbicides could selectively kill some plants—hopefully the weeds—while others—hopefully the crop plants—could live another day.

For farmers who had witnessed hand hoeing by gangs of laborers—paid or forced—the sight of weeds curling up and dying while their crops stayed green must have seemed unbelievable. All they had to do was spray the field and weeds like beggarweed would be gone.

Unfortunately, selectivity also meant that a herbicide could selectively kill one group of weeds and allow others to live. Some herbicides killed only small-seeded weeds and left large-seeded ones to grow. Some killed only broadleaf weeds and left grasses. Farmers responded by mixing several herbicides together to kill different types of weeds at the same time. The trick was to do this without injuring the crop.

For peanut farmers, finding the right mix of herbicides was challenging because the crop was easily injured by the chemicals. It was especially difficult to kill weeds that were closely related to the crop. Peanuts and beggarweed do not look alike or act alike, but their metabolism is similar. A herbicide that killed beggarweed usually injured peanuts. A herbicide that was safe to use on peanuts was generally useless against beggarweed. Farmers might apply one herbicide to control grasses, another to kill small-seeded broadleaf weeds, but still some beggarweeds would survive. And with those other weeds out of the way, the beggarweed survivors thrived.[60]

By the 1960s, peanut farmers came to rely on herbicides as their main method of weed control. For Florida beggarweed, they relied on dinoseb herbicide, the one I heard them talking about at the Southern Oven Diner.[61] Dinoseb smelled bad and came in oily metal drums, but farmers sprayed it on hundreds of thousands of acres every year. Dinoseb was great when it worked; it selectively killed most early emerging weeds, including many beggarweed seedlings. But it didn't work well if the soil was too wet or too dry. Beggarweed could sneak by, germinating late and escaping control. With resilience and human assistance, beggarweed had escaped to Africa and North America. With that same resilience and chemical selectivity, it escaped control efforts to become the most common and troublesome weed in peanut fields.

What farmers couldn't have seen in the early days of chemical agriculture was that the drudgery of pulling and chopping weeds all day in the sun would be replaced by struggles of a different sort. Reliance on these miracle products would yoke them to the whims of a powerful chemical industry. Farmers would find that the only solution to the failure of one herbicide was to spray yet another herbicide. When it came to weeds, nobody gave much thought to possible dangers to health or the environment. Nobody gave much thought to farm consolidation, a different sort of selectivity. The weeds had to go. Farmers wanted more options for getting rid of them. When herbicides came along, the options they wanted were more herbicides.

That's where things stood when I got to Georgia in 1983.

The Ultimate Rise and Tumble

One morning in early October 1986, the phone was ringing when I got to work. I figured it was my technician calling to check in. We had so many studies going on, we could work for days without seeing each other: experiments on seed germination, morphological variation, weed competition, and biological control. We were inching toward an Integrated Pest Management (IPM) approach to reduce beggarweed emergence and growth and apply herbicides only if needed.

Instead, the call was from a colleague in another state. In the storm of colorful words, I understood that the Environmental Protection Agency (EPA) had abruptly halted all use of dinoseb herbicide. The decision was immediate and final—no hearings, no rebuttal. Dinoseb had been used since the late 1940s. It had been approved based on safety data from a well-known laboratory. Farmers from Georgia to Texas, Oklahoma, and New Mexico sprayed dinoseb.[62] We needed an explanation.

What we didn't learn until much later was that eighty drums of dinoseb herbicide had plunged over the side of a Danish ship on a January morning somewhere in the stormy North Sea. With engines knocked out, the ship drifted into port. Nobody knew where the cargo had tumbled. The Danish environment minister insisted the drums be found. This prompted a German company that manufactured dinoseb to run new toxicology tests. The results were startling: dinoseb increased risks of birth defects in female field workers; it could cause sterility in men. Fish, birds, invertebrates, and mammals were also at risk. Moreover, dinoseb could persist in wells. It was later revealed that dinoseb's alleged safety was based on data from Industrial Bio-Test Laboratory, which had submitted many flawed reports.[63]

Peanut farmers, peanut butter manufacturers, peanut sales reps, peanut haulers, peanut dryers, peanut roasters, peanut handlers, peanut processors, peanut marketers, peanut market speculators, peanut product promoters, and peanut commodity commission folks were irate and bewildered. Their first target was the EPA. Farmers had already purchased supplies of dinoseb for the next year. It was sitting in drums out behind barns across the South, waiting to be mixed and sprayed. Applying that dinoseb was now illegal. If dinoseb residues were found in somebody's field, their entire harvest would be destroyed.

The second target was state and USDA agriculture experts—folks like me—who were supposed to be solving the problem of weeds in peanuts. Now, more than ever, farmers wanted more options, more tools. Every state in the peanut growing region had at least two or three people working on weeds in this crop. Didn't they have some options? Why did some bureaucrat in Denmark—where dinoseb was not even used—know more about the chemical than anybody on this side of the Atlantic? And what about all those drums of dinoseb waiting to rust and leak?

The growing season was just a few months away, and the $1.03 billion peanut industry asked how they were supposed to get by on 1.5 million acres full of weeds. Farmers would not be interested in my yet-to-be-developed IPM approach, the studies on beggarweed seed germination, morphological variation, or competition. My colleague CW asked me if the biological control fungus I was studying could be mixed with chemicals to control beggarweed. I didn't know if he was joking; he was not inclined to bad jokes.

I sat in my office and tried to figure out how to respond. Outside the window, crop stubble from the previous season covered a grey landscape. The calls were relentless. Farmers hinted there was no telling what would be sprayed on fields come spring. Confection companies predicted lawsuits due to pieces of weeds in their peanut candies. Herbicide companies assured me their product would replace dinoseb. I met with farmer groups, commodity organizations, product development groups, snake-oil salesmen. Without the dinoseb, weeds would not be controlled, the price of peanut products would rise, foreign competitors would take advantage of this, farmers might get desperate, and somebody might get hurt. What would we do about Florida beggarweed? Did I have any clever IPM tools to handle that?

The next few years were the height of Florida beggarweed glory. More than anything plant evolution could accomplish, and beyond what promotions and seed sales promised, the overdependence on one tool along with gross negligence toward others spirited the scraggy dryland shrub to extraordinary ecological success.[64] Without dinoseb to knock it back, a beggarweed renaissance festival filled every peanut field. In springtime, the soil was covered with the grey-green seedlings. You could spot them from miles away.

Farmers dragged out old cultivators and pulled them across fields to remove the weeds between the rows. Inevitably the cultivator shovels injured peanut branches, which lie flat on the soil. Dirt thrown onto peanut stems by the cultivator promoted infections of white mold and other diseases. By mid-season, bushy beggarweed, thick as a privet hedge, intercepted fungicide applications and made disease control impossible. The weeds cut out light the peanut plants needed to produce a good yield. When weather turned dry, beggarweed, which had evolved in a desertlike environment, grew for weeks without water. Irrigation systems spurted out across the landscape to keep the peanut crop alive, and beggarweed simply grew taller, with more leaves. Enough beggarweed seeds were produced over a few seasons to keep the soil supplied with beggarweed seedlings for a generation.

Farmers did their best, but by the end of the growing season, so many beggarweed plants survived that it was impossible to see the crop beneath the tangled forest of sticky, hairy stems. Some sent large crews of migrant workers into fields to pull up beggarweed plants. It was expensive, but it made the difference between a harvestable crop and a tangle of weeds not worth the trouble to dig. Others tried to mow down the tall plants. But still the stems got caught in the chains of peanut diggers. The plants didn't invert properly so they didn't dry properly. When harvesters rattled across the fields, the hairy, resin-covered beggarweed stems wrapped around the forward reel, plugged the auger, gummed up the cutter bar, and got caught in the threshing cylinder.

Farmers wanted options right now, so they could get the production machine working again. But all my work had been shaped by a time scale that had little value in the face of urgency. The industry turned to the guys who sprayed and prayed. I was happy to give them the attention. Eventually, good old CW came up with a chemical concoction that was as almost as good as dinoseb, if it was applied right, and a whole lot safer. Over time, other products came on the market that were even better. These new tools brought the glory days of beggarweed to an end. Populations started to decline. Fewer seeds were dispersed, and fewer seedlings emerged in fields.

Farmers were thrilled. For months, I would wonder if I had made the right decision to focus on management approaches that might be useful ten or twenty years down the road. If farmers had been able to use IPM methods, we might not have gotten into that mess, but the research was incomplete. Instead, industry pushed farmers into the same approach they had used before, relying on one herbicide sprayed year after year. The result was predictable: selectivity favored one group of weeds over others.[65] The decline of beggarweed had little to do with the research I had conducted. It was due to new herbicides that opened the way for other weeds that became the new menace on the landscape, and left Florida beggarweed to grow in their shadow.

Before I left Tifton to take a job back in Ohio, I drove across the muggy back roads one last time. It was late July and crops were at their peak. Some folks said you could hear them growing. Florida beggarweed was still out there in the peanut fields and a few other crops. But nothing like before. I wondered if anything could have been done, many years ago, to keep that weed from getting into these fields in the first place. Or maybe its presence was inescapable. If there was going to be agriculture, there were going to be weeds. If not beggarweed, some other species like the ones already taking its place that some other weeds guy would get a chance to tangle with.

The view across the fields was impressive. The agricultural machine sucked up sunlight, water, and petroleum products to pump out three thousand pounds of dry nuts per acre. A few abandoned center-pivot irrigation rigs said things were changing. But farmers were determined to keep the machine churning. The industry, university, and government agriculture experts were stuck in it, too, like beggarweed seeds in your socks. So was anybody who bought a jar of peanut butter from a grocery shelf.

Lingering and Complicated Spirit

I've been back to Africa several times in the last few years. I recently took a walk on the edge of a farm field in Tanzania, past tomatoes and stalks of papayas. My host identified weed species and translated the local names. I always have an eye out for my favorite weeds. Eventually, I spotted a patch of Florida beggarweed off in a corner.

I could only speculate how it got there. But I knew something of its path to Africa and beyond. Beggarweed's unintentional dispersal was not unlike other plant species. What made Florida beggarweed unique was its path to extraordinary weediness in a single cropping system. It was not a neglected genotype of a cultivated crop, like wild oats or weedy rice. It was not genetically close to a major crop. It became a useful forage crop on its own merits, an accident of human history. Seed dispersal aided by human activity is common for weedy plants. But segmented seed pods with clutching, hooked trichomes are not. Florida beggarweed was unique in its dual personality. It had potential to supply quality forage, to improve soil, and—with appropriate human manipulation—to be a troublesome, noxious weed.

Tanzanian farmers were using it the way Kobe and I might have used it in Ghana years before, had he been willing to tolerate the itching. Here it was a forage crop to feed animals and a green manure crop to improve the soil. In this setting it is not a weed. It is being evaluated in this way in India, East Asia, Egypt, and around the Mediterranean.[66] People see it as a species with potential to be useful. As good coevolutionary partners, they are selecting genotypes that grow more favorably and increase the ecological success of the species. Wherever people tangle with it, the species will spread to new habitats. In some of those habitats it will settle down again as a weed. It is no longer a dominant weed in the American South, but it infests soybean fields in Brazil. With ongoing dispersal, along with climate chaos, Florida beggarweed, which can survive in hot, dry, unpredictable conditions, is not through with its journey toward ecological success.

At the village I visited in Tanzania, the locals referred to beggarweed by a name common in Africa. It translates roughly to "Spirit plant," a tribute to its wildness and ability to show up in unexpected places. In the United States, the Department of Agriculture recently renamed it "Dixie ticktrefoil," unintentionally, perhaps, bearing a trace of its complicated American journey and path to weediness.

Velvetleaf, *Abutilon theophrasti*.

3

VELVETLEAF

George Washington was famously fastidious about his clothes. In April 1789, he stood on New York's Federal Hall balcony as the new leader in a role that he alone had the power and obligation to shape. How should a president elected by the people—the wealthy, landholding, white, male people—behave? How should he look and dress? The old general moved stiffly among ladies in European finery and men in London-tailored suits. No one could have failed to notice that the Commander wore a plain brown coat and breeches, the product of Hartford Woolen Manufactory of Connecticut.[1] He had told Lafayette, "It will not be a great while before it will be unfashionable for a gentleman to appear in any other dress." His inaugural statement was brief, awkward, and forgettable. The lasting message was Washington's attire: to be truly free and independent, the new nation cannot rely on materials whose supply is subject to the whims of other nations. It must build its own industry, beginning with textiles. Fashionable or not, this is what national sovereignty looks like.

Although flax and wool industries had begun to make clothing, other textile products such as bags, twine, and rugs still came almost entirely through English ports. Most important were textiles linked to the navy and shipping industry, which still relied on the mother country for sails, nets, and—significantly for the future weedy flora of American agriculture—all manner of cordage, or rope.

In the early days of the Republic, rope had the same significance for domestic security that oil, rubber, and rare earth elements have today. Rope was a strategic commodity, essential for economic activity and national defense. Fully rigged ships might have fifteen or more sails edged with boltrope and raised with lift cords. There were bowlines, clewlines, buntlines, leech lines, jibs, and shrouds and more.[2] Large fighting ships, like the USS *Constitution*, carried more than 4.8 kilometers (3 miles) of cordage weighing around 54 metric tons (60 tons). It had several anchor cables the diameter of a stove pipe. The whole economy was tied together by rope: vessels for transport and trade, ships to protect transportation and trade, and ships to impose the "national interest" in other lands.

Washington needed a way to encourage the growing of fiber plants and the manufacture of cordage. When his dull brown suit failed to perk up the local industry, he and Alexander Hamilton devised a tariff. The tariff would raise the price of imported goods and encourage production by Americans. It was also a way for the underfunded federal government to raise money for things like self-defense, including building and outfitting a cordage-dependent navy.

The first tariff act put duties on imported goods that Americans already produced locally, like nails, iron, and glass products. One of the principal objects of this and subsequent tariffs was a duty on imports of all types of cordage and crops from which cordage could be made. The raw materials for making cordage, rope, netting, twine, and similar items were derived from a group of mostly unrelated plants. In the trade, all were classed as "hemp."

Hemp was a generic word for several crops, their fibers, and fiber materials. The plant whose fate was most tightly twisted with "true" hemp (*Cannabis sativa*), was a tall, fibrous herb that became known as velvetleaf (*Abutilon theophrasti*). The colonies depended on Russian hemp imported on British ships. Sometimes, velvetleaf was an accidental contaminant, sometimes a purposeful adulterant, in those shipments. "Hemp" might mean either plant, or the two of them together.

The resulting confusion is important for the story of how velvetleaf became a major weed. Government or industry actions to safeguard the manufacture of cordage, or the growing and processing of strategically important hemp, advanced the adaptation and spread of velvetleaf. With his signature on the Tariff of 1789, the Father of our Country set in motion efforts by the federal government to protect and promote hemp—a future illicit drug—and velvetleaf—a future noxious weed and scourge of agriculture.

Tangled Identity

Velvetleaf belongs to the mallow family, Malvaceae to botanists. This name might not ring a bell, but everyone knows of the clan that includes okra, hibiscus, hollyhock, and cacao, the source of all things chocolate. The Malvaceae also boasts four types of fiber crops: several species of cotton, two of jute, another called kenaf, in addition to velvetleaf. Most are adapted to warm, humid environments, and have spread throughout most of the world. The origin of the *Abutilon* genus is unclear. Some say India, others say around the Mediterranean.[3] The Russian biogeographer Nikolai Vavílov pointed to northern China, because of the many varieties and ancient uses there.[4]

Early human ancestors discovered the essentiality of rope. Long, thin stretches of animal skin, vines, or strands of grass could pull, carry, catch, and wrap food gathered and hunted. Rope was used for bow strings, sacks, and garments. It held those fig leaves in place so discretely. Grids of rope made nets, traps, and bridges. With rope came pulleys and mechanical devices to move, lift, and build structures in ways that stretched the imagination and forever altered the environment. Like any new technology, the invention of rope changed society. When used by some to gain greater access to resources, others were compelled to join, or were left behind.

Velvetleaf fibers were used in China before the Zhou dynasty (ca. 1027–256 BCE). Farmers made velvetleaf fiber twine, shoes, and itchy, burlap sacklike clothing. Growers selected velvetleaf seeds from plants with tall stems and carried them throughout China to barter or share. Velvetleaf seeds moved as contaminants in plantings of cotton and other crops. Over centuries, velvetleaf plants and seeds traveled from China through Russia and south toward the Mediterranean.

The mixing of velvetleaf with hemp began in Central Asia. Hemp was also an ancient fiber crop, but the two plants are not related. They don't look alike, but both are tall annuals that thrive in close plantings to produce straight stems with long silky fibers. Hemp fibers are strong and smooth, but the plant is finicky and can be hard to get started. Velvetleaf fibers are short and rough, but plants are easy to grow. Velvetleaf filled in the spotty stands of hemp. Farmers harvested and processed both plants together and fiber users hardly knew the difference.

Velvetleaf mingled with hemp in fields, fibers, and languages for many centuries. The Chinese name for velvetleaf is *qǐng má*. The character *má* signifies "hemp," and has long been used generically for fiber plants and textile fabrics. Hemp and velvetleaf, along with jute, ramie, and other fiber plants would all be called má or have "—má" as a final syllable.[5] In trading ports and markets, few could distinguish velvetleaf from the more desirable true hemp, which fetched a higher price. Consequently, velvetleaf became "Abutilon hemp" or "China hemp."[6]

The mingling of velvetleaf with hemp that begun in Asia was a prelude to the muddling of names and identity in the hands of Western botanists. Linnaeus called it *Sida abutilon*.[7] That was soon changed to *Abutilon avicennae*, to honor the great Persian physician Abu Ali Sina, who had coined the name *Abutilon*, meaning a mallowlike plant.[8] But precedent had been set four years earlier, with the name *Abutilon theophrasti,* to honor the Greek physician, Theophrastus. Assigning a single genus-species tagline was supposed to bring order to plant nomenclature, but at least a dozen Latinized names still clutter the scientific literature. Common names are just as messy.

Uncertainty about the name and identity aided the plant's dispersal and persistence into modern times. Until people agreed on a name, its nature remained unclear—is it "butterprint," the lovely plant we keep in the garden for the seed capsules that make decorative imprints in butter? Or is it "rattle box," the pernicious weed we've been trying to destroy?

It's hard to know why this was so difficult, since the plant is quite easy to recognize and distinguish from almost any other species, including others in the mallow family.

New World Curiosity

There are no records of purposeful shipments of velvetleaf from Western Europe to North America before colonial times. A plant resembling velvetleaf was in the hands of the Flemish physician and botanist Rembert Dodoens in the 1500s. An English version of his classic *Cruydeboeck* includes an illustration of a plant labeled "Yellow mallow." The depiction more or less fits velvetleaf, such as the heart-shaped leaf, buttery flower, chambered capsule, and general structure—"three or four cubits high."[9]

Velvetleaf seeds traveled to North America along with the Russian hemp seeds delivered by London trading companies to colonial farmers. Seeds also came attached to velvetleaf stems mixed in loads of raw hemp bound for rope makers. Seeds may have contaminated early lots of soybean seeds sent from China's port of Tientsin, near where velvetleaf was grown for fiber.[10]

Early colonial records of a plant resembling velvetleaf are suspect. A 1680 description from Virginia by John Baptist Banister is vague and was likely copied from elsewhere.[11] John Clayton's 1743 flora made note of a plant with a "small yellow flower and strong smelling leaves."[12] Clayton and Banister might have been the source for the Englishman Philip Miller's *Dictionary*, which described it as "very common in Virginia, and most of the other parts of America."[13] Later writers would point to Miller's book as evidence that velvetleaf was widespread before the Revolution. This is unlikely. Miller never saw the plant, and his book repeats taxonomic and factual errors of other books. Moreover, some of the best early American botanists, like the Bartrams, Rafinesque, and Nuttall, did not mention a species resembling velvetleaf in accounts of their plant explorations in the 1700s. Neither did the English botanist John Josselyn nor the Finnish naturalist Peter Kalm, who wrote credibly about explorations of New England.

The first convincing description of velvetleaf in America was that of Benjamin Smith Barton in 1803. He was a friend of Jefferson and adviser to Lewis and Clark. Barton described "indian mallow" as an annual plant, with wooly, heart-shaped leaves, and yellow flowers, found from Vermont to Virginia and west to Albany, New York.[14]

Barton was also credited with resolving a pressing biological mystery of his day. Linnaeus, who never walked through a North American forest, had the impression that American plants were *"peculiarly* smooth" (his emphasis). He understood that Native American people were similarly without hairs. However, after extensive study, Barton announced to the scientific world: "We now know, that the Indians of America are not more smooth than are the Japanese, the Chinese . . . and many other nations." Barton was reassuring: Linnaeus would find "in our woods, very many species covered over with all the various kinds of pubes," including, specifically, "Sida Abutilon," his name for velvetleaf.[15]

For a while, many botanists considered "Abutilon hemp" to be native to North America. The plant is tall, dusty green, with large, velvety heart-shaped leaves, and seed capsules like a squat okra. It resembles no other plant of eastern North America. Nevertheless, the question of its origin had conflicting implications. Some biologists believed that European plants and animals ranked higher in stature than those from the wilds of North America. An Abutilon of civilized European extraction would be expected to behave as a crop. If native to the new and untamed continent, it would be dismissed as unimportant, along with other native inhabitants. Thomas Jefferson was one who believed it was native to America.[16] Henry Muhlenberg's 1793 list of more than a thousand plant species from eastern Pennsylvania included "Indian mallow . . . abutilon," along with notes suggesting it grew wild and might be native.[17]

The word *Indian* throughout American history has not made things easier for botanists. "Indian mallow" could be a plant from the continent of India, or a plant long used by native people of the Americas. The word *Indian* also had the connotation of "false," as in Indian tobacco (*Lobelia inflata*) and Indian hemp (*Apocynum cannabinum*), signifying botanical imposters. When a respectable botanist like Muhlenberg used the name "Indian mallow," it could mean he believed the plant was native, which it was not; that the plant was introduced from India, which it was not; or that it was some sort of false mallow, which it was not.

One thing it was not called in colonial days, was "velvetleaf." Another thing it was not called, was a "weed."

New Twists on Sovereignty

The United States was jolted by the Napoleonic wars of the early 1800s. Americans were particularly insulted when the Royal Navy forced US sailors into service on British warships. President Jefferson responded with a series of embargoes, high import duties, and a last-ditch attempt to cut America off from the world with a trade policy of "nonintercourse." The policy, at the end of his term in office, had no lasting impact except for the stir it caused in high school history classes for the next two centuries.

The prospect of another conflict with England woke Americans to how dependent they remained on foreign goods needed to prosecute another war. Much of the fighting would be on water. Ships and their supplies took on renewed significance. President James Madison pleaded with Congress for measures to boost manufacturing of essential materials for the navy and shipping industry: "A thriving agriculture [must produce] substitutes for supplies heretofore obtained by foreign commerce."[18] Opportunists in the Northeast launched manufacturing to support the shipping industry and to profit from the war. Farmers expanded plantings of all types of hemp.

When smoke finally cleared over the Battle of New Orleans at the end of the War of 1812, Madison hailed agriculture and industry as a "source of national independence and wealth."[19] The message was as clear as Washington's broadcloth suit: never again shall the nation's sovereignty be threatened by dependence on foreign rope.

The chaos of war was over, and James Monroe was installed (1817) as Madison's successor. Farmers were poised for progress in agricultural innovation and diversification. The country slowly transformed from an undeveloped frontier to an urbanized industrial empire.[20]

Within two years, overextension of credit threw the nation into its first major depression. Monroe reached for a familiar economic tool, the tariff. Duties were placed on foreign goods, including hemp and cordage. For the next decade, protectionists pushed for higher tariffs to stimulate domestic production. The most protective tariff, the infamous 1828 "Tariff of Abominations," put the highest duties ever on hemp, along with other essentials like molasses, indigo, and distilled liquors.[21]

There was one peculiar problem with a tariff to encourage US hemp production: American hemp wasn't very good. It couldn't match the quality of imported Russian hemp. Russian laborers dried hemp stalks carefully on raised trellises, but American farmers let the stems fall behind the scythe to dry on the ground like hay. It got rained on and mildewed. The secretary of the navy deplored rope made from American hemp as weak, flimsy, and unsafe for use on battleships.[22] Abutilon hemp was of no help. Its shorter fibers made rope that was weaker and harder to handle. It was an impurity, an imposter, no substitute for the real thing. Sailors on the lookout for enemy ships were not disposed to risk their lives scaling an eighteen-meter (sixty-feet) high mast on ropes of Abutilon hemp. They would get high only on good Russian hemp.

The association with true hemp had helped Abutilon assimilate in North America. But times were changing. In the early 1830s, Robert McCormick patented a "hemp brake" that mechanized the separation of hemp fibers from the stalk and improved the quality of rope. But it didn't work for Abutilon fibers that had to be pulled from stems after a week-long soaking. The new hemp brake technology, not protective tariffs, ensured domestic fiber independence. Farmers expanded hemp production in Kentucky, even when import duties were reduced with the 1832 tariff, never to rise again. Other innovations followed, including seed cleaners that removed contaminants, like velvetleaf seeds, from hemp seed lots. This ended the agricultural and linguistic entanglement, and velvetleaf could no longer benefit from confusion with hemp.

It was just as well: ship-building materials and technology were changing, too. The boom in Kentucky hemp production got busted when iron chains replaced shipping rope. Coal-powered steamships began to cross the Atlantic. They didn't need strong winds and large crews of sailors. And they didn't need miles of rope, rigging, and cable.[23]

Abutilon hemp's prospects for ecological success unraveled. Fields of mixed hemp and velvetleaf were replaced with permanent pastures, where annual weeds persisted only as seeds in the soil. No longer tied to hemp, Abutilon joined abandoned crops like indigo, sand-grass, goosefoot, and other plants of marginal worth.

From Lags to Riches

William Darlington, of West Chester, Pennsylvania, was a remarkably talented guy: physician, merchant, traveler, soldier, congressional representative, library founder, bank president, canal commissioner, railroad administrator, and more. Still, he took time to botanize along back roads not far from Washington's Valley Forge retreat. In 1826, Darlington published a book on the flowering plants of the area. He described "Sida abutilon . . . Indian Mallow, De Witt weed, Velvet-leaf, a naturalized foreigner, and becoming troublesome in our cultivated grounds."[24] The source of the name "De Witt weed" is a mystery, as it occurs in no other source. But this is where "velvet-leaf" makes the first appearance I have been able to find. Darlington couldn't have been the first to notice the stems and leaves covered by dense, fine, velvety pubescence. Nor could he have been the first to hold a weedy perspective on the plant out in the countryside.

When he wrote the first book on North American weeds, in 1865, Darlington's view of velvetleaf (no longer De Witt weed) had not improved. He now considered it "a worthless and troublesome intruder—frequent in Indian-corn fields, potato patches, and other cultivated lots—and is of a size sufficient to be a nuisance. It should be always carefully eradicated before it matures its seeds."[25] Harsh words from a guy who named a plant after its short hairs. Anyway, the farm press picked up on this sentiment, calling it, "quite a pest. It delights in rich soil, and on the cultivated prairie lands attains a great size."[26]

It's not surprising that it took one hundred or more years for farmers to recognize velvetleaf's potential weediness in a habitat where it had not evolved. This pattern is common to organisms introduced to a new range. Successful establishment is not guaranteed, whether it be a plant, insect, fungus, or virus. Many—probably most—don't make it. Velvetleaf survived the initial introduction because, for a while, human companions tended the seedlings along with their hemp crops. When left to grow feral, the orphaned velvetleaf population went through a lag phase, where growth in number and expansion to new areas were relatively slow. The lag phase can be a few years or centuries until people recognize it for what it is: a pest spreading into new and unexpected places.

For velvetleaf, the lag phase was a time of intense selection for geno-types that suited its new status as an abandoned would-be crop. The plant is mostly self-pollinated, but there was already substantial genetic varia-tion in the population due to its history of introduction from Asia. Genes encoding traits that allowed velvetleaf to succeed as an untamed species increased in frequency. Having been cultivated as a crop in Asia, velvetleaf retained genes for traits that were useful in field crop environments. But genes for traits that farmers had favored—like uniform growth or low seed dormancy—were no longer useful. Velvetleaf abandoned, or dedo-mesticated, crop traits and reacquired traits that would help it succeed as a weed.[27]

Darlington observed velvetleaf population expansion beyond roadsides and waste areas into crop fields. He recognized specific features that al-lowed it to persist and spread wherever soil was disturbed, and crops were grown.[28] Two features that allowed velvetleaf to be more than just a roadside ephemeral or garden escapee are seed dormancy and growth plasticity.

Plugs, Plastic, and Propagule Pressure

The marvel of velvetleaf seeds lies in their ability to sit and do nothing for years on end. Velvetleaf seeds can remain dormant, but viable, in the soil for more than fifty years. A characteristic of most successful weeds is that they find some way to put seeds into a dormant state and release them at just the right time to coincide with crop planting. Imagine waiting patiently for fifty cold winters, wet springs, hot summers, and chilly au-tumns. Then one time the soil is turned, a seed finds itself in the right place with warming temperatures and springtime rains. The seed awakens, ger-minates, and a new seedling appears.

Velvetleaf seed dormancy is relatively simple: a hard seed coat prevents water from entering the seed. Inside, the seed is fully mature and ready to go, but water can't get in to start germination. It is sealed off by a layer of cells just beneath the seed coat. At the spot where the seed was once attached inside the capsule is a little plug, or cap. When the plug loos-ens, water enters, enzymes are activated, cells expand, and a swelling root pushes out of the seed. Once the plug is loose, there's no going back to a dormant state. If the seed doesn't germinate, fungi and bacteria get inside and destroy it. As students in my lab decreed: "Germinate or die."

But velvetleaf is too clever to behave quite so predictably. Not all of the seeds stay dormant for fifty years. Only a few do. Most pop their plug and germinate at some random time over the years in response to freezing and thawing during springtime. Soil movement during tillage can scratch seed coats, allowing water in to begin germination. And a few seeds delay germination to later in the season, too late for farmers to get equipment into the field to control them.[29]

The other key weedy feature of velvetleaf is plasticity: the ability of a plant to change shape, branching, height, or other characteristics in response to changes in the environment. A plant can sense its environment and activate or deactivate genes for physiological pathways that allow it to adjust to conditions or to stress.

Plasticity allows velvetleaf to change its height and branching. Having been selected as a crop for its long stem fibers, velvetleaf retained an unusual ability to adjust stem growth to its surroundings. When surrounded by short plants, velvetleaf readily pushes up above them, branching out and stealing sunlight to make more leaves, flowers, and seeds. The same velvetleaf plant growing in a field surrounded by tall plants will stretch the central stem farther upward to reach over top of the canopy. After completing that upward growth, it branches out to catch the sunlight. This type of plasticity allows velvetleaf to compete effectively with short and tall crops to capture resources from the environment and turn those resources into seeds that have a chance to do the same the next year.[30] Or for the next fifty years.

Darlington recorded changes in the flora over about sixty years. His shifting concern about velvetleaf followed its expansion. Eventually, the aggregate velvetleaf population—the metapopulation—reached a tipping point where sufficient propagule pressure led to explosive exponential growth. The exponential growth phase is a period of nearly unlimited growth in density and extent. Once velvetleaf began exponential growth, it was impossible to turn back the biological clock to the innocent days when it was just a misunderstood curiosity.

The rapid expansion of velvetleaf populations in North America in the late 1800s made possible its dominance in agronomic crop fields through the next century. Land was cleared for planting, competitor species were removed, and velvetleaf spread along with crop production. In farm fields, velvetleaf plants would compete with the crop and each other for sunlight, water, and soil nutrients. This imposed intense agrestal selection that enhanced adaptation to managed agricultural settings.

When grown as a crop since ancient times, velvetleaf had been consciously and unconsciously selected for desirable crop traits, reducing the frequency of weedy genes (traits) in the population. When left to go wild, velvetleaf had less use for crop traits, and weedy traits were favored. Velvetleaf plants that got a head start, grew faster, reached taller, produced more leaf area, and were able to reproduce more quickly had an advantage. Agrestal selection favored these genotypes and made them more competitive, persistent, and successful as a weed. Whole fields became covered with velvetleaf; the type and timing of soil disturbance made it a particular weed in fields where corn was grown.

But no plant can get around the fundamental laws of biology. The constant final yield law limits the production of weed biomass (total weight of plant matter) in a given area. As the density of plants increases, biomass also increases—but only up to a point. Eventually, resources become limiting. As the density continues to increase, adding more plants causes more intense competition for resources and total biomass no longer increases. It's a zero-sum game: the addition of another plant is balanced by the decline in growth of others and eventually the net change in biomass is zero. The more there are, the less each one can grow. The fewer there are, the more each one grows.

This law worked in velvetleaf's favor. It doesn't require thousands of them to produce a lot of velvetleaf biomass. A relatively low density of individuals with space to stretch out will do the trick. Within an area of, say, one square meter, a single velvetleaf plant without competition can produce as many stems, leaves, and seeds as one hundred or one thousand velvetleaf plants fighting for the resources in that same area. This makes it difficult for farmers to shrink the velvetleaf population once it becomes established in a field. Reduce density by 90 percent, and the remaining 10 percent can potentially cover the soil with as many leaves and return as many seeds as the 90 percent that the farmer labored to remove. The theory, math, limitations, and conditions for measuring the constant final yield are more complicated than the common sense understanding of it. In the end, the weeds—many or few—will always win.[31]

Meeting of Opportunists

Velvetleaf topped the agenda at the monthly meeting of Philadelphia's Franklin Institute on October 16, 1862. Patent attorney Henry Howson delivered a paper on the fiber properties of "American jute." The topic generated great interest in the local textile industry, which supplied fabric to clothe the Union army. None of the plants he discussed was actually related to true jute (*Corchorus* spp). After describing fibers of the local perennial, *Hibiscus moscheutos* (rose mallow), he moved on to "the *Abutilon Avicennce*, an annual, readily cultivated, and hitherto considered a useless weed. The fibres of this plant . . . are of a silky character and extraordinary strength." This is one of the first instances where the crop and weed character were acknowledged, and the weed character mostly ignored in the face of hope and praise for the crop. Howson droned on: "A company—of which the proprietors of the patent will be members— is about being organized for the . . . planting . . . and the preparation of the fibres for the market. The members present will be satisfied that success must attend the efforts of an enterprising company who follow up with energy the prosecution of this important invention or discovery."[32]

Out West, in Hannibal, Missouri, Henry Draper was equally hopeful. He had discovered a new material for making fabrics, cordage, and twine: "The lint of the plant called by botanists *Abutilon avicennae* . . . having various local names, as 'mormon weed,' 'stamp weed,' etc., and is found growing wild and in great profusion in Missouri, Illinois, and other states." Draper patented his process, remarkably similar to that used for hemp, and assigned it to James McConnell, of Springfield, Illinois, who offered one hundred dollars for a ton of Abutilon fibers to feed a processing plant he hoped to build.[33]

McConnell hawked his abutilon thread, rope, and twine at the 1871 Illinois State Fair. He encouraged farmers to adopt the plant that "grows so freely upon any rich soil, even thrusting itself in and growing spontaneously, that it has almost come to be considered a farm pest in many portions of the country."[34] As though to encourage that outcome, he provided instructions for sowing 160,000 to 220,000 seeds per hectare (65,000 to 90,000 per acre). He assured farmers that the crop would not exhaust the soil, so long as the leaves, branches, and seedpods were returned to the field.

Unfortunately for Howson in Philadelphia, an "American jute" company never materialized. Nor did McConnell's rope and twine exhibit convince Illinois farmers to plow up corn fields and plant velvetleaf for an imaginary industry. Boosting velvetleaf's prospects throughout America's richest agricultural landscape would require something more than simple entrepreneurial enthusiasm.

Mission Impushable

Financial panic that began in Europe in 1873 collided with the economic upheaval following the American Civil War. By 1877, the United States sank into another depression. Labor turmoil festered in eastern factories and railroads, while agrarian radicalism simmered in the West.

In New Jersey, legislators reached into the nation's history for a reliable solution to economic insecurity. The State Board of Agriculture warned that in addition to massive unemployment, poverty, and deprivation, the nation was dangerously dependent on countries like India for $7 million annually in imports of jute. Once again, American sovereignty was threatened by dependence on foreign fibers.

The board proposed, "the introduction of new industries . . . to extend and diversify agricultural pursuits."[35] New fiber crops would put farmers back to work and supply raw materials for manufacturing. New industries would create jobs for the unemployed men in the cities and exploit a whole new generation of women and children in textile mills. They would develop three crops: silk (from mulberry trees), ramie (the tropical plant *Boehmeria nivea*), and jute—not true jute, but "American jute"—velvetleaf.[36]

Mr. Samuel C. Brown, secretary of the New Jersey Bureau of Labor, Statistics, and Industries, disseminated a pamphlet, "by order of the Governor," to encourage farmers to plant these crops throughout the state. Brown had grown them in his backyard and was therefore confident of their suitability. Farmers were asked to gather seeds of American jute, put them in an envelope, and send them to him for some undescribed evaluation. He advised farmers to add manure to the soil, sow the seeds, and watch them grow. There was no time to waste: American jute was sprouting up in fields from Ohio to Illinois; other opportunists threatened to tie up the market.

To strengthen his case, Brown inserted into the report several pieces of evidence. First, was a letter from Professor Sylvester Waterhouse of Washington University in St. Louis, verifying that the American plant "has the spirit and capacity of conquest. With invasive march it has taken possession of large tracts of land. Its tenacity of life and rapid spread render its cultivation a far easier task than extermination. . . . To the manner [*sic*] born, it exhibits a stubborn determination to occupy its heritage."[37]

This was followed by a six-page contribution from Emile LeFranc, president of the Southern Ramie Association. LeFranc reported results of his studies on American jute, which,

> convinced me beyond the least doubt of the possible development of these fibrous productions, towards large avenues of trade and wealth for New Jersey. . . . American jute is an excellent fibrous plant. . . . As it prospers best in manured spots, it cannot be called properly a wild plant. But until its utilization shall be established, it will be a noxious weed which farmers consider and treat as a pest. . . . Every year the promiscuously sown seed extend the growth of the plant to the great inconvenience of farmers. But it evidently has an irrepressible mission to fulfill.[38]

In case this didn't convince the board to abandon this scheme, further testimony was provided by J. L. Douglass and Daniel F. Tompkins of the Special Committee of the New Jersey State Fair. "American jute . . . has long been considered a great nuisance to the cultivator of the soil. It grows in all the parts of New Jersey in a wild state . . . as a common weed." The State Fair, convinced that the plant could be grown economically as a successful fiber crop, awarded American jute the "Highest honors of the Society—a gold decoration of merit."[39]

In the year following the report, nobody volunteered to plant more of the already common mulberry trees for laborious silk-making. Nobody bothered to plant ramie, the tropical crop that would not grow in New Jersey. Nobody wasted effort on true jute, which had already failed in several southern states. That left only one option to reignite the economy and once again assert America's sovereignty in the dynamic market for twine and burlap bags.

Excitement about the solution to the economic crisis spread westward. The Chicago *Daily Inter Ocean* reported that something called the "Devil's Plant" and known to botanists as "Abutilon avicennae," could produce a fiber, "infinitely superior to Indian jute." The plant was familiar in the Midwest as "a pest, and a bad one at that. It makes its appearance . . . here, there, and everywhere, and no device known to the farmers will rid them of it . . . growing until it is head and shoulders above the tallest of tall men. . . . Yet. . . . it has proved to be a devil in disguise."[40] After a few more clichés, the article suggested that given its new status, perhaps a more propitious name was in order. Soon, experiments like those conducted in New Jersey were initiated to test American jute, hemp, and ramie production for new farming ventures in Oregon.[41]

In a bold move to encourage its spread and adaptation to a wider range of soils and environments, to build up a soil seedbank that would persist for decades, and to further the ecological success of a weed that farmers would battle for the next century, the New Jersey Senate passed an Act in 1881, to encourage the production of velvetleaf. For every ton of three-foot-long "American jute" stems, farmers would receive a bounty of five dollars. News of the initiative was announced in newspapers throughout the state. It was left for farmers to respond.[42]

For some reason, a gold star and five dollars did not entice farmers to hack through acres of velvetleaf stalks in order to spawn a new line of textile manufacturing. No velvetleaf stems were delivered. Eventually, the bounty appropriation was rescinded. Samuel C. Brown later lamented that a "lack of interest and spirit of cooperation . . . and well-nigh universal apathy exists [among farmers] in respect to the introduction of new industries."[43] Alternatively, New Jersey farmers understood something about the nature of velvetleaf.

Efforts to industrialize velvetleaf got caught in the push-pull conundrum facing the commercial development of a new crop. Velvetleaf was easy to grow, and its use as a fiber crop was ancient. New Jersey farmers didn't deliver tons of stalks because there was no factory sitting idle for want of them. Factories didn't exist because there was no efficient mechanical method to remove the fibers. And no American company would invest in expensive velvetleaf fiber extraction machines when the competition was pennies-a-day labor in India's jute industry. And besides, how many burlap bags and balls of twine do we really need?

Farming lore is full of promising new crops pushed into markets that weren't pulling them. Practically all plants (especially weeds) contain chemicals with properties that could potentially be exploited in commercial products. Velvetleaf was long known for its medicinal properties to reduce swellings and to clear the eyes.[44] The seeds have been pressed for oil, which contains the antifungal constituent hibicuslide-C.[45] Over the last fifty years, velvetleaf has been mentioned in patents for textiles, paper, health foods, rod handles, hearing-loss medicine, and preparations for skin bleaching and hair-growing. A weed might be a plant whose virtues have yet to be discovered. But most, like velvetleaf, remain weeds in spite of well-known virtues. There just isn't enough demand for things like velvetleaf skin cream to persuade a farmer to fill crop fields with more velvetleaf seeds. As a result, velvetleaf has, in modern agriculture, never been much more than just a weed.

V Is for Victory—and Velvetleaf

While opportunists promoted velvetleaf from New Jersey to Oregon, the landscape for velvetleaf's ultimate rise to weedy success was being prepared in the Mississippi Valley. European-style agriculture had spread with land clearing, wildlife killing, and forest burning throughout the nineteenth century. Farmers tested all kinds of crops to find those that would grow best and spawn new industry. Cotton, rice, and tobacco were planted into Ohio and New England before being relegated to the South, where rice, sisal, pineapples, and other crops had failed. The corn-beans-squash system of the indigenous people had long been reduced to mostly corn, with a few other crops that served the needs of cities, factories, and schools, all fed by one of the richest farming systems on the planet.

By the mid-1800s, second-generation immigrants of mostly European background came to consider themselves the rightful occupiers of the land. This entitled them to push native inhabitants off the prairie and to exclude others outside the "myth-symbol complex of the 'American' ethnic."[46] These ethnic Americans pushed most of the native grassland ecosystems off the landscape and replaced them with their Eurasian crops, weeds, animals, and pests. This transformed the landscape into an alien agricultural system that became accepted as the rightful use of the land.

"Where once the prairie stretched along the banks of the river," remarked
Illinois historian Frederick Gerhard in 1857, "stately mansions of modern
civilization may now be found—and near them many a foreign plant,"
among them, "the marsh-mallow of Vincennes, Abutilon avicennae."
Gerhard's understanding of plant ecology was equally poetic: "Unlike
their human prototypes, these plants. . . . stand peaceably side by side,
deriving their nourishment from the same parent, imbibing the due of the
heavens, and enjoying the light equally diffused over them, of the glorious
sun of Diety."[47]

This perspective inspired a new agricultural industrial complex whose
rightful place seemed to be in the heart of America. It was spurred, at
the start of the new century, by America's dependence on Manchurian
shipments of oil and meal derived from soybeans. In 1928, a group of
Illinois farmers, processors, and farm organizations agreed to buy, at a set
price, all the soybeans produced from fifty thousand acres. The next season, farmers delivered enough soybeans to operate a mill at Bloomington,
Illinois. Extraction yielded two viable products: soybean oil and soybean
seed-meal. Farmers suddenly had a market for every bushel of soybeans
they could produce.[48] In 1928, around 220,000 metric tons (8 million
bushels) of soybeans were produced on 234,000 hectares (579,000 acres)
of land. During the next ninety years, production would increase by a
factor of 550.

The spectacular rise of the American soybean industry has no parallel
in agricultural history. It happened without tariffs or bounties. Soybeans
had languished as an Asian curiosity during colonial times, and a marginal
hay crop into the 1940s. Suddenly, it was the leading commodity crop
in North America, pulled along by a combination of changing markets,
demographics, diets, and lifestyles where vegetable oils, rather than lard,
were increasingly in demand.

The rise of velvetleaf as a major weed followed soybean's success.
It was pulled to a level of prominence and infamy that few weeds have
achieved. By the 1940s, velvetleaf was recognized as a weed of row crops,
especially corn, where its remarkable plasticity allowed it to tower above
the crop canopy. Farmers could run a cultivator between corn rows four
or more times, and strong velvetleaf stems would hold up firmly, especially
those growing within the corn rows. As soybeans became more common,
velvetleaf found ideal conditions in which to thrive. Soybean plants are

not nearly as tall as corn plants, so velvetleaf didn't have to grow so tall to reach the light. The resources saved could be put into producing more seeds. Velvetleaf and soybeans came from similar regions in China; they were already adapted to the same growing conditions. They germinated about the same time, flowered in unison, and matured their seeds simultaneously. When harvesting machines gathered soybean seeds in the fall, they also gathered velvetleaf seeds and spit them out with crop residues across the soil surface.

Velvetleaf followed in the smoke trails of soybeans when World War II created critical shortages of domestic oils and fats. American conflict with Japan blocked US access to about a billion pounds of East Asian oils, including soy. Like Washington and Madison before him, President Franklin Roosevelt needed a way to promote a domestic industry that would reassert US sovereignty. Tariffs were out of the question, and one can hardly imagine what kind of promo Roosevelt would do on behalf of vegetable oils.

Instead, the government encouraged soybean production with a pamphlet entitled, "Soybean Oil and the War: Grow More Soybeans for Victory." The final line emboldened farmers: "When you grow more soybeans, you are helping America to destroy the enemies of freedom."[49] It didn't mention that when they grow more soybeans, they make more land suitable for velvetleaf.

Soybean plantings rose rapidly, from 2.4 to 4 million hectares (6–10 million acres). The government instituted a price support system to make this possible. It offered to deliver harvest machinery, which incidentally spread weed seeds with the chaff. With this appeal, North America became the world's premiere soybean producer and exporter. As Asia disintegrated into chaos, the Manchurian market for soybean oil was replaced by US farmers, who started supplying oils and fats to the Allies. Butter production declined and was soon replaced by margarine, made with soybean oil.

Escape to Freedom

When the prospects for velvetleaf expansion and productivity could not have been more promising, the great talents of academia, government, and industry intervened at the end of the war, with the invention of the herbicide 2,4-D. This part of the story is familiar. We saw it in dandelion (*Taraxacum officinale*), an early and enduring target of 2,4-D. We saw it again in Florida beggarweed (*Desmodium tortuosum*), under attack from dinoseb herbicide in southern peanut fields. A similar pattern involving other herbicides are part of the success stories of many other weeds. But velvetleaf was never the target of a single herbicide. It was the missed target of *every* major broadleaf herbicide. Velvetleaf was more troublesome in soybeans than in corn. Soybean acreage was expanding at a rate that would soon exceed that of corn. So the chemical industry set its sights on finding a soybean-safe herbicide to kill velvetleaf.

The adoption of new farm technologies follows the same sort of pattern as the invasion of a new weed into the landscape. The initial introduction is followed by a lag period while a few early adopters take a risk on a small scale. Others follow, and at some point, the technology becomes widespread. But 2,4-D was an exception. When 2,4-D became available in the late 1940s, farmers were so desperate to kill weeds, there was no lag period for adoption. It was cheap, effective, and satisfying to see weeds like velvetleaf curling, crisping, and dead. It was hard to resist. If the neighbor used it, they'd grow more corn, make more money, and buy more land—maybe yours if you didn't do the same.

But 2,4-D could not be used in soybeans. It kills broadleaf crops (e.g., soybeans) as well as broadleaf weeds. The first herbicides that could be used safely on soybeans killed mostly grasses and only a few small-seeded broadleaf weeds. That left large-seeded weeds like velvetleaf to escape, free of competition from other species. Any velvetleaf genotypes injured by the chemicals were eliminated from the population, leaving only non-susceptible genotypes to survive and reproduce, another type of agrestal selection.

By the time atrazine herbicide was widely available for use in corn (around 1960), grain farmers, with the help of industry marketing efforts, began to think of herbicides as their main approach to weed control. Farmers' reliance on herbicides ignited the American free enterprise system in a race to spray every acre of corn and soybeans with the next new miracle product. Big money would be made from a herbicide that killed grass weeds in a grass crop like corn. Bigger money would be made from a herbicide that killed broadleaf weeds—like velvetleaf—in the broadleaf crop soybeans.

In 1969, the Monsanto Company introduced a new preemergence herbicide called alachlor, to control grass and broadleaf weeds in corn and soybeans. It was highly effective, highly promoted, and widely adopted. Deep in the world of plant spirits, it might have been possible to hear Abutilonian cries of joy, for velvetleaf is almost completely immune to alachlor. All the annual grasses and most of the other broadleaf weeds were killed or suppressed by alachlor. Velvetleaf jumped from the soil and towered over the soybean crop with little competition from others. In corn, the combination of atrazine and alachlor became the standard treatment. Atrazine would control 60 to 80 percent of the velvetleaf plants, but those remaining were genotypes with higher tolerance of the treatment. And besides—remember the constant final yield law?—the survivors, even at a low density, could potentially produce as much velvetleaf biomass—and as many seeds—as if all the plants had lived.

For the next thirty years, atrazine and alachlor became fundamental to weed control programs in the corn-soy Midwest. By the 1990s, they were the most commonly applied pesticides in US agriculture. They were also the most commonly found in groundwater.[50] Alachlor, being more soluble than atrazine, moves readily into the soil and along channels into groundwater from which drinking water is drawn. To be clear, not all 32 million kilograms (70 million pounds) of herbicide applied annually drains into wells and rivers. The movement and degradation of herbicides in the environment depends on soil and environmental conditions. Most of the herbicide molecules break apart along the way, some binding to soil, and some metabolized by microorganisms. Still, in the 1980s and 1990s, samples of surface and ground water had levels of atrazine and alachlor higher than health advisory limits. The long-term health effects of exposure to alachlor were uncertain.[51]

With industry consolidation, farms got bigger and farmers fewer. Labor and fuel costs spiked, and farmers substituted more herbicides in place of mechanical cultivation. Farmers began to plant soybeans in narrow rows to get faster soil cover. This made mechanical cultivation impossible and increased the dependence on herbicides. The herbicides worked well. Just not well enough.[52]

After a few years of using a particular herbicide mix, some weed species escaped the toxic brew, velvetleaf among them. The logical response was to apply more and different herbicides. Weed control recommendations included four or more herbicides mixed in a single cocktail. The result was good control of most weeds, leaving only the hardest-to-kill species. Farmers simply sprayed higher rates or more chemicals. Many thought they had no alternative, especially those who adopted reduced-tillage practices and no longer owned a cultivator. American agriculture was the embodiment of free enterprise, but farmers found themselves increasingly stuck on a herbicide treadmill with no way to jump off.

Anxiety's Miracle Cure

I returned to Ohio to conduct research on weeds in the late 1980s. Herbicides that killed velvetleaf were at the center of social anxieties about agriculture. The pesticides were being found in Ohio waterways. They were killing honeybees. The memory of the disaster in Bhopal, India, was still fresh, where a herbicide manufacturing plant had blown up in December 1984, killing more than thirty-seven hundred people. In and out of the farming community, concerns were raised about the environmental consequences of herbicides sprayed on fields of corn and soybeans.

The single-minded reliance on mixtures of herbicides to reduce velvetleaf populations was not working. It was impossible to kill all the velvetleaf plants. Those that survived produced enough seeds to maintain a population of seedlings year after year.

My research group—graduate students and research assistants—approached the situation with the simple observation that velvetleaf was not (yet) taking over the world. Something was keeping it in check, even where farmers didn't apply herbicides to kill it. Like all weeds, velvetleaf lives in a Malthusian world. Air, water, nutrients, and other resources are limited. Plants compete for them. Predators and pathogens abound. Exponential population growth has limits. Our research aimed to figure out what limited velvetleaf populations. Maybe we could make use of whatever was already at work constraining velvetleaf's desire for greater growth.

Over the next ten years, we conducted experiments on velvetleaf competition, seed production, seed survival, predators, and parasites. We focused especially on seeds, the key to velvetleaf's survival. We counted, spread, buried, and exhumed thousands of them. An old sewing machine clanked for hours in the lab to make small bags of mesh fabric. We counted out 50 seeds per bag and buried them at different depths. Farmers driving by asked what we were doing and left shaking their heads: only a university weeds guy would think it was a good idea to plant velvetleaf seeds in farm fields. We retrieved the bags throughout the year and developed tricks to induce the seeds to germinate. Insects that might be predators of flowers or seed capsules were put to the test. Ground beetles, slugs, and other animals that ate velvetleaf seeds off the soil surface were monitored and their damage quantified.

It wasn't so hard to find constraints on velvetleaf populations. In spite of its tough seed coat, rapid growth, phenotypic plasticity, and competitiveness, velvetleaf was not invulnerable. It didn't take over the world because it was attacked by naturally occurring pathogens and predators. Stems were invaded by insect larvae; flower buds were eaten by bugs; and seeds succumbed to fungal attack.[53] Velvetleaf turned out to be a nitrogen addict; it nearly withered without high fertility. Seed production could be greatly suppressed by competition from tall crops managed properly. The seeds were easy prey to field mice and birds. We had to figure out how to encourage these restraints on population growth so farmers could slow down the herbicide treadmill.

I approached funding agencies, industry, and crop commodity organizations about the idea of exploiting natural constraints on velvetleaf populations. The rationale seemed compelling: newspapers were full of reports of herbicide contamination and cancer rates among farmers. And we had a sufficiently outlandish idea: we would put pieces of the puzzle together on a field scale to see if the impact of velvetleaf could be reduced and allow farmers to rely less on the most environmentally harmful herbicides.

After being turned down a few times I got a message from a proposal review panel in the mid-1990s: there was no need to study biological constraints on velvetleaf or other weeds. Agricultural chemical companies had recently introduced modified genes to make crops resistant to glyphosate (Roundup) herbicide. Farmers would soon be planting the GMO crop seeds, spraying glyphosate over all their fields, and killing all their weeds, velvetleaf included. With GMO technology, there was little interest in other approaches, no matter how clever.

I had long learned to not take such rejections personally. This one was instructive in a different way. Something about weeds in the human consciousness can persuade agricultural experts that profit-driven technology operates outside the laws of population biology and basic genetics. Something about weeds removes doubt about the success a new technology, its public acceptance, the wisdom of spreading it across the whole Midwest, the consequences for rural communities, and its impact on what agriculture could and should be. Research on natural constraints was headed the wrong way against a moving train. Skepticism was in short supply.

Around the Next Corner

When glyphosate-tolerant GMO crops were first used, farmers expected their weeds to be wiped out. I had been working closely with David, a farmer who looked forward to weed-free fields of GMO corn and soybeans. Like most farmers, David knew that if he used the same weed-killing chemical for several years, any weeds that survived would slowly take over. In some cases, weed populations had shifted to entirely new groups of species. Each time, the survivors were harder to kill. David already had weeds that survived his old atrazine and alachlor program. Spraying glyphosate on GMO crop fields was supposed to get those weeds under control.

He gave me a call when he saw some worrisome escapes. I stopped by his farm the next day along with my colleague, Ted Webster. David wanted us to look at a patch of velvetleaf poking through the soybean crop. He said there were rumors that maybe glyphosate didn't kill all the weeds. There were stories that some weeds that had been easy to kill were getting harder to control. Is velvetleaf going to be one of those weeds? If so, he would be in big trouble—it would spread across the farm and he'd have no other chemical to control it.

Ted looked at the pattern of infestation across the field—regular, linear stripes. Probably just a herbicide application error. For all we knew, velvetleaf was easily killed by glyphosate. The escapes were just a mistake.

David was relieved but not convinced. He had heard about a patch of velvetleaf plants down in Knox County. The farmer sprayed it with Roundup, and they didn't die. He sprayed again, and still they didn't die. Somehow, they escaped. A guy from Monsanto showed up, pulled them all out, put them in bags, and took them away. That's what he'd heard.

Next afternoon, I called the local county extension office and spoke to a colleague with long experience scouting weeds. He knew Knox County, and he'd also heard the velvetleaf escape story. It was supposed to be down Route 3 and off County Road 321. You go around a sharp curve. There's a big white barn on the corner. And a white fence. You'd run right into it if you didn't make the turn. Velvetleaf had been growing right there, near the barn. The soil was still turned up from where the plants had been pulled. That's what he'd heard.

I threw shovels and bags in the truck and picked up some marking pens. I asked Ted to come along. He would be more thorough and more careful with the sampling. By the time we got to Route 3, we had experiments planned. The plants were gone, but we would find seeds in the soil. Nobody knew better than us how to extract velvetleaf seeds from soil and coax them to germinate. We'd grow them out in different conditions and try to figure out what was going on. Then we'd follow up with studies on possible genetic variants. The next five years of research were just waiting for us, in front of a barn, somewhere along these winding roads.

Knox County is named after Henry Knox, Washington's secretary of war. The county rests on the edge of the Allegheny Plateau that was left after the last glaciation. Probably nobody knows how many sharp curves lie among the rolling hills and valleys. Or how many big white barns guard the peaks of those curves. And how many white fences. We drove the back roads in the lengthening shadows until we couldn't make out scratches in the earth. Finally, around another sharp curve, a farmer backed a hay binder into a big white barn behind a white fence. We slowed to a stop and peered into the darkness for signs of disturbed ground. The tractor engine died, and huge doors slammed shut.

No, he hadn't heard about a patch of velvetleaf. He called it buttonweed. Sure, there was some on his farm. But he killed it pretty good this year. It was worse last year. Maybe all that rain. His grandpa said he never saw it on this farm before his son took over and started using chemicals. Well, you couldn't farm without chemicals the way things were nowadays.

But anyway, he had it and wished he didn't. It was a mystery how a weed like that—velvet-weed, button-weed, or whatever you want to call it—got there in the first place. The air turned cool and cicadas sang in the woods. We stood in the dark for a while, swatting mosquitoes. There wasn't much more to say.

Hope in a Jar

In 2012, archeologists recovered a plain storage jar from a Late Neolithic site in southern Hungary. In the jar were 934 seeds that are unmistakably velvetleaf.[54] I wrote to the dig team leader, offering expertise in nursing velvetleaf seeds to life, and a collection of velvetleaf genotypes for comparison, to help piece together their story. A kind reply assured me that after fire and seven thousand years there was no hope the charred seeds could germinate. The rest of the story remains buried.

Velvetleaf seeds seven thousand years old. In a jar. The seeds had been intentionally harvested, cleaned, and stored. They may have been used for food, medicine, or for planting next year's crop. They might have been collected from random fields, or a tended garden site. Nobody knows. Seeds like these had moved across continents. Some were consciously selected, carried, traded, or shared. Others, in their dry capsules, moved with water, eroding soil, or were blown across the crust of snow-covered fields.

With a different twist of history, those seeds might have been ancestors of a highly useful crop, one that brought prosperity to farmers and supported a successful fiber, medicine, or skin cream industry. It would have become an economic crop in modern agriculture after many years of conscious, human-mediated, selection. Different from the agrestal selection that made it a weed. The result would be a narrowing of the genetic base, discarding weedy traits like seed dormancy, plasticity, and plant-produced chemicals or superfluous hairs that provide protection.

A velvetleaf crop would have phenomenal uniformity, just as you see in any corn or soybean field across the Midwest. Every plant would be the same height, with little branching, emerging from the soil on cue, flowering simultaneously, setting seeds in unison, maturing and drying down just in time for harvesting machines to cross the field and collect commercially useful plant tissues of uniform size, shape, color, and dryness. Industry would have selected velvetleaf genotypes with high productivity, suited to fertile and coddled environments. As a botanical handmaiden, velvetleaf's photosynthetic machinery would be diverted to the most marketable plant parts. And there is little doubt we would now see fields of genetically modified velvetleaf plants with novel genes coding for proteins that protect against disease, increase yield, and, of course, stand up to particular herbicides.

The germplasm of this hypothetical velvetleaf crop would be owned by some company—in fact, all of velvetleaf germplasm would likely be owned by a single company. Velvetleaf varieties would be created that follow the rules set by industrial agriculture. There would be a Velvetleaf Growers Association with offices in state capitals and Washington, to manage a checkoff program and lobby for legislation that protects velvetleaf grower interests and increases consumer demand in export markets. And on occasion, to advocate for tariff duties on Chinese velvetleaf imports.

Would this be greater ecological success?

To the extent that it had a choice in the matter, velvetleaf took the weedy path. If we can abide such agency, we'd see that it became attached to its coevolutionary human partner by providing practical fiber material in order to gain transport from northern China to the turned prairie sod of North America by way of rich farmland in Russia and Eastern Europe. It inspired the imagination of lawyers, farmers, and entrepreneurs whose expectations propelled its dispersal to new environments. With every new farming practice, agrestal selection improved its adaptation to diverse field environments and tolerance of weed control practices. But it was careful not to lose the traits that allowed it to escape, persist, and survive. The more widely it spread, the more its genetic potential continued to be expressed as favorable alleles increased in the population.

Some writers have suggested that crop plants essentially induced humans to facilitate their ecological success by domesticating them. They argue that humans selected plants with useful traits, protected them from pests, and spread their genes to habitats across the globe. But I wonder: if those domesticated industrialized, human-protected species had a sense of themselves as biological creatures, would they consider themselves more successful than their weedy counterparts?

Velvetleaf resisted domestication, maintained its weedy genetic base, adaptability, and variability. It doesn't follow anybody's rules. It doesn't depend on any company or foreign country for its survival and continued adaptation. From the view of velvetleaf, it has achieved a sort of botanical sovereignty.

Yellow nutsedge, *Cyperus esculentus*.

4

NUTSEDGE

Evening had fallen, and a few lights flickered from wires twisted around the grass-roofed shelter. Subtle rhythms of Tanzania's *ngoma* music drifted in the background. I sat before several rows of farmers. They looked me over. Many wore sweaters in the cool air that rolled off the lush mountains between Lakes Rukwa and Malawi. Most had spent the day tending fields of maize or harvesting tea or cocoa. The local agricultural extension officer introduced me with customary respect as the outsider.

I began with a Kiswahili greeting, hoping to soften skeptical eyes. I explained that I wished to talk about weeds; not to lecture, but to listen and learn. I asked basic things: What makes a plant a weed, for you? What weeds trouble you the most? In what ways do they trouble you?

The first responses were formal and polite. Like most farmers, they were eager to talk about weeds. They knew a lot about them. They relaxed into gentle disagreement about what to name them. Even in unfamiliar languages I understood a resigned frustration.

After a time, a woman with deep set eyes stood up: "Mr. John," she began, "we cannot afford to hire labor. We must use family labor. Family labor means the wife. The wife must do the weeding. We take young ones to the field. They sit by a tree. We work in the sun." Quick translations to Bungu and Kinga met with general approval. "But still, I must cook. I must clean. I must tend to children. The elders. But how, if I must spend my day rogueing these weeds?" Nodding heads, whispers, and shifting in seats. "These weeds, they are killing us."

Soft murmurs, a quip from the back row, and muted laughter opened a rambling discussion. All agreed that labor for weeding was in short supply. Some farmers hired outsiders—people whom others did not know or trust—to weed the fields. Some sent young children to the field instead of school. Older children who finished secondary school took jobs in the city, far from tedious farm work. They didn't care to inherit a farm and spend their days weeding.

By this time, nearly everyone was engaged, and I struggled to follow the conversation. I understood that pesticides were being promoted to save the labor of hoeing. Dealers contracted with men—husbands or others—to spray the poisons on fields where the women and children worked. The women were not happy about that either. It was expensive, unfamiliar, and contrary to how families had farmed for decades. The effect on their health and the health of their children was unknown. Some of the neighbors sold out to big farmers with tractors and sprayers and noisy machines. Foreigners bought up large tracts of land and put family farmers out of business. Some villagers had quit farming to become vendors, buying cheap goods in the city and reselling them in the central market.

In the melee of languages, I picked out a word, *nati-nyasi*, repeated several times. The extension officer leaned over to explain: the Kiswahili word meaning "nut-grass." Yes, of course. It didn't matter what crop the farmers were growing. In the tropics, yellow nutsedge (*Cyperus esculentus*) was bound to be the weed known to every farmer. I had heard it in many languages. And I had heard similar stories years before, from women in West Africa who spent hours bent over, a baby on their back and a short-handled hoe in their hands. The same dilemma confronts smallholder farmers throughout Africa, as well as Asia and Latin America. Laborers—mostly women—spent days hoeing and pulling weeds like yellow nutsedge. If not, their crops couldn't grow, all the work of preparing soil, planting seeds, protecting the crop from insects and diseases, and harvesting would be for naught. It was all because of weeds.

Getting at the Root

Earlier that year, I had responded to a request for volunteers to provide technical expertise on weeds and weed management for a project in Tanzania. The world of international development has a changing jumble of acronymic aid agencies, including an alphabet soup of NGOs (nongovernmental organizations). This "Farmer-to-Farmer" project was a collaboration between two consortia, managed in-country by a third consultancy. The aim was to provide "education and stewardship training" in maize and peanuts. Reading through the official Work Plan, it seemed that the real goal was to increase the "safe and effective" adoption of herbicide spraying on these crops. I was willing to volunteer because I had experience in Africa, knowledge of maize and peanut production, and skepticism about the use of herbicides in smallholder farming.

I landed in Dar es Salaam and was greeted by the NGO project supervisor. He said new policy directives had changed the objectives. I would not focus on peanuts, a crop I knew well, but on sorghum, about which I knew little. Also, I would not do education and training; I was to design demonstration experiments to test different weed management methods. He introduced the local representative on the project, Mr. V, a short, round, bespectacled entomologist. An expert on stem-boring insects, Mr. V seemed curious that I, or anyone, would take more than a passing interest in weeds. I was not clear if he was following an agenda or expecting me to direct our activity, so he could politely object. The next day, after formal meetings with officials, greetings from their bosses, and handshakes with the boss's boss, Mr. V and I were handed a revision of the revised work plan. We were to focus on maize, vegetables, and rice. We would not design experiments. We were to find places to locate those experiments. Mr. V smiled and held up a GPS device to record the coordinates.

Our driver was a linguistically versatile guy named Marra. His family were seminomadic Maasai herders in the Rift Valley, a life that he left to become a driver and tour guide. On our first drive into the city, Marra bolted from a stoplight, headed toward a gap in oncoming traffic, and turned sharply onto another highway. When my blood pressure approached normal, I ventured that I was not in a hurry; a safe return was my main goal. Marra caught my eye in the mirror: "No worries, Mr. John. I am a safe driver. You will see."

Over the next few weeks, we traveled over thirty-five hundred kilometers (twenty-one hundred miles) south to north across Tanzania. Marra would slip a recording of smooth jazz into the player, and from the seat of a Land Rover, I saw a vast, beautiful country. We passed through brush and thicket-spotted savannas, moist "miombo" woodlands, acacia-dotted grasslands, baobab forests, snow-capped mountains, and pitch dry plains. It was late October, the short rains were late, a good time to view giraffes, water buffalo, zebras, elephants, impalas, and other animals across the leafless landscape. I couldn't bear to photograph them in their embarrassingly emaciated condition. Marra knew their biology and behavior, explaining where each stood in the social order.

We turned off the main road toward villages with no electricity or running water. Vibrant towns came alive at night by kerosene lanterns and candlelight. People were moving, loading, hauling, hammering, tinkering, chopping, cooking, and never too busy to stop along the street, exchange a greeting, ask after the family, touch hands a third time, and give their blessings. A visitor to these villages would never go hungry or lack a place to sleep. This was the Africa I had come to know years before. Here, the workdays were long. People walked everywhere. Life wasn't easy. Yet there was a sense of community, regard for honest labor, and respect for the work of human hands.

We bounced along rutted paths and grassy fields looking for locations for demonstration sites. At every stop, farmers gathered under a mango tree, an extension worker appeared, and we talked about weeds. Marra, who seemed to know every local language, stood at the back and provided translations. The stories varied by crop and region, but in every village the concerns and weeds were similar. Women from the southern maize-growing region expressed displeasure at their situation: the men prepared the land, planted the seeds, and left the women to do the rest—weeding and more weeding. There it was again, nati-nyasi. The men took the harvest, and the money that came from it. Men in the crowd did not contradict them. Elsewhere, we met vegetable growers who farmed and marketed together, selling their produce along the main road. They said labor was exchanged and scheduled among the group, although I saw only women hoeing onions with their narrow, short-handled hoes.

The farmers had few options for removing weeds. Most used manual labor—their own or that of their spouse, their children, or cheap hired help. Their fields had many kinds of weeds, but always nutsedge. They'd pull or hoe the nutsedge plants, but the weed quickly regrew from underground tubers. Some farmers had turned to herbicides—expensive, poorly understood toxins that are out of reach for the poorest farmers. Applied correctly, at the right rates, they could save time and labor without harm to the crop or environment. Applied at the wrong time, to the wrong crop, at the wrong growth stage, at the wrong rate, with the wrong equipment, the wrong calibration, or poor water quality, and results were wasteful, damaging, or worse.[1]

Farmers in every village seemed to agree that weeding was the least desirable activity. Insects and plant diseases were problems in some crops, and in some years. Farmers knew how to manage them. But weeds like nutsedge came into every crop in every season. To farm meant to hoe weeds. Weeds, and the labor to pull them, were part of the background. As a weeds guy it was gratifying to meet people who knew and cared about weeds. But it left me uncomfortable, dreading the time someone might ask me to solve their dilemma.

The Thin Green Line

If any plant was born to be a weed, it was yellow nutsedge. Its success comes from being unlike any other weed. It looks like a grass with its long, narrow leaves, and is commonly called "nut-grass." But nutsedge isn't a grass and doesn't behave like one. Nor is it considered a broadleaf weed for these same reasons. It likes moist soils, but is not a rush, like those gathered by the water. As the name implies, yellow nutsedge is in a different category—a sedge—a heat-, sun-, and water-loving plant that most gardeners and farmers would recognize as nothing but an intrusive weed.

Along with its cousin, purple nutsedge (*C. rotundus*), the nutsedges represent a syndrome of weediness. They've been branded the world's worst weeds.[2] Although purple nutsedge is more difficult to control, yellow nutsedge is more widely adapted, geographically distributed, and culturally significant.[3] You're as likely to find yellow nutsedge in tropical sugarcane plantations as in Alaskan cabbage fields. The two species are somewhat different in morphology and behavior, but both are aptly named for the underground "nuts" (tubers) on which they rely for dispersal and survival. While much of the biology pertains to both species, I will focus on yellow nutsedge to be true to the facts of this particular story.

Nutsedge belongs to the Cyperaceae (cy-per-AE-cee-ee), a family of more than five thousand species that evolved in humid tropics. Plants in the *Cyperus* genus have been used for oils, foods, and materials. *Cyperus papyrus* is the source of papyrus.[4] The weedy nutsedges probably originated in tropical Africa.[5] Their sleek stem is three-sided, triangular in cross-section. The stem arises from leaves clustered at the soil surface, one leaf on each side of the stem. At the top is a flower that resembles loosely tangled bottle brushes. The flowers are straw yellow, and the rest of the plant is yellow-green and sweetly scented. Nutsedge is a perennial, but it behaves like an annual in its pattern of emergence and flowering. In temperate regions, the entire plant dies back at the end of the growing season, leaving only underground parts to carry on.

Though born to be a weed, nutsedge could also have been born to be a crop. There are two morphs. Weedy types are "nutsedge," the noxious plant that causes havoc in farms and gardens worldwide. Crop types are "chufa" (also "tiger nut"), cultivated for the juice extracted from their tubers. The edible chufa varieties have larger, sweeter tubers (chufas) than the weedy nutsedge type. The two types diverged genetically, but not enough to be separate species—they still cross-pollinate. Their coevolutionary partners have led them down different paths that have selected for tubers of different shape and size. Chufa types were nursed, coddled, harvested, and thought to be a crop, eaten raw, boiled, roasted, or pressed for their juice. Nutsedge types were cursed and attacked with all manner of mechanical devices and are considered a globally despised weed. Botanically, they are the very same species.

Which came first, nutsedge or chufa? The answer isn't clear. Chufa is an ancient crop, but nutsedge, strangely, might not be an ancient weed. Chufa was domesticated in ancient Egypt.[6] The name *esculentus* is an old word meaning "edible." Theophrastus wrote of chufa tubers boiled in beer and eaten as a sweet treat around 300 BCE. A thirty-five-hundred-year-old painting on an Egyptian tomb shows workers tending the nuts; beside the picture is a recipe for mixing chufas with honey.[7]

But ancient, tasty chufa is an unlikely source of weedy cast-off nutsedge. Chufa types rarely produce flowers, so they could not have generated genetic variants that became weedy. Nutsedge types do make flowers. Their viable pollen moves in the wind to cross-pollinate with other nutsedge plants and produce viable seeds. So only the weedy nutsedge could have generated genetic variants with different morphologies and tuber sizes. Chufa types were domesticated long ago and have remained stable varieties. Where cultivated, they haven't bred weedy biotypes. Yet, while Theophrastus, Pliny, and Columella knew and wrote a lot about weeds—and all enjoyed chufa snacks—none mentioned a weed resembling yellow nutsedge.[8]

A plant known and cherished from ancient times could not spawn the world's worst weed without the help of human accomplices. Their path to weediness began long ago in North Africa, where people grew and selected plants with large tubers for planting in gardens. Plants that devoted energy to tuber growth, put little into making flowers. Any plants with small tubers were discarded in junk heaps, where they proliferated as weeds. Junk plants put energy into making flowers that yielded many seeds and many genetic variants. The tiny seeds and small tubers moved widely with the help of humans and animals. By twelve thousand or more years ago, weedy nutsedge had become adapted to tropical and temperate climates. Chufa types had traded genetic variation for human affection, and as a result did not move far from their subtropical homes.

Arab merchants likely passed both types from Egypt, south into Africa, and westward across the Mediterranean toward Spain.[9] Weedy nutsedge found its way into agricultural systems on nearly every continent. Efforts to hoe, chop, pull, and torment yellow nutsedge removed the weaker genotypes from the population and increased the proportion of those that could tolerate hoeing, chopping, pulling, and tormenting. This unwitting agrestal selection for robust genotypes in agricultural environments led to many locally adapted weedy nutsedge variants. The weedy types, which still produced a respectable tuber, were sometimes mingled with crop types, unintentionally or not. In good growing conditions, tubers of weedy nutsedge can attain a size and sweetness similar to the cultivated chufas.

The species that split to become nutsedge and chufa is not the only plant with a dual personality. Carrot, parsnip, radish, lettuce, amaranth, okra, rice, millets, and other crops all have weedy counterparts—the same genus and species—almost identical genetically. In turn, some plants generally recognized as weeds, including purslane, nightshades, pigweeds, and chicory, have varieties that are used as crops. The crop types were selected to behave as accommodating, cooperative, and useful species. The weedy types, like yellow nutsedge, misbehave as obnoxious, competitive pests that put a drain on human labor and the economy. The line dividing them is thin.

Things Lost in the Background

Midway through my East African weed tour, I visited an area of expansive upland and flood-irrigated rice production, a cropping system new to me. Square fields surrounded by irrigation ditches were plowed by animal traction or gasoline-powered tillers. After harvest, large canvas mats were stretched on the ground for rice to dry before being loaded into burlap sacks. Marra drove us to a large supplier of agricultural machinery, fertilizers, and pesticides. The proprietor eagerly showed off his modern facilities. A steady flow of overstuffed pickup trucks and donkey carts carried sacks of seeds, crates of rice hulls, and unmarked plastic barrels in and out of the warehouse.

From there we moved down the supply chain to visit the middlemen (sometimes middlewomen), called "agro-dealers." They operated small shops or kiosks stocked with a smelly assortment of pesticides, seeds, and fertilizers, along with backpack sprayers, rubber boots, and crates of soda and bottled water. Colorful plastic bags, white-capped bottles, and silver-topped cans lined shelves behind a glass counter. There seemed to be two or three agro-dealers in every village. Most had about the same inventory. They sold weed killers with product names I didn't recognize, but the labels revealed a short list of mostly older herbicides, like 2,4-D and propanil. Sometimes I couldn't make out the contents because the labels were in Chinese or, rarely, Kiswahili.

Some shops appeared to be well run. The proprietors knew their products and how to use them. More often, the shops were staffed by youngsters, boys of about fourteen, who seemed to have little idea what the products on the shelves were used for. After seeing this at several dealers, I got curious about these kids, out of school, shy, uncomfortable in the presence of an outsider.

At an agro-dealer near the mountains I tried to engage the young man behind the counter. His name, he said, was Alfred. He lived in the next village. The shop belonged to his uncle. I asked what was in a green and yellow container, and he simply shrugged. Along a lower shelf, were unlabeled bottles that once held soda or water. These, I knew, had been filled with leftovers from spray tanks. I asked about one with a milky white liquid, and Alfred handed me a bag of fungicide used on vegetable crops and rice.

I pointed to several unlabeled glass bottles with what looked like Coke inside. Alfred motioned toward a jug on another shelf. The label was in Mandarin, but he knew the name: "Gramox," he smiled, paraquat herbicide. I asked if he could tell me about it. He looked at Marra and they spoke an unfamiliar language. The farmers know what they want, Marra explained, they come with their money, and Alfred gives it to them.

The next day was bristling hot. We drove to a patchwork of rice paddies that stretched to the foothills of Mt. Kilimanjaro. The paddies were all the same size, the soil mounded along the edges, and water flowing in the irrigation canal. In the corner of some paddies was a raised platform. From this vantage, a young boy spun a leather sling to whirl small balls of clay with remarkable speed and accuracy to scare birds away from the ripening grain. A dozen or so farmers gathered to greet me and talk about weeds. The subject held no amusement for them. With the heat and constant water, there was no stopping the continuous germination and growth of weeds. Farmers used a herbicide to kill grasses and small-seeded broadleaf weeds. But this just cleared the way for other weeds more difficult to control. The men did the spraying, and women followed a few days later to pull weeds that survived.

In the crowd, I recognized Miss Ruth, a serious, willowy woman, one of the agro-dealers we had visited. She was eager to show me her rice fields, the cleanest I had seen. Hers was a uniformly lush stretch of rice. The throbbing sun reflected in the water and snow-capped Mt. Kilimanjaro stood in the background. Miss Ruth walked the edge of the field, stepped into a patch of rice, and yanked a plant from the mucky soil. She held it out: nati-nyasi. She spoke rapidly, raising three bent fingers; leaves arranged in threes needed no translation. She pointed to the ground and held the tip of her little finger between sun-beaten thumb and forefinger, describing the tuber. Reaching down again, she pulled from the irrigation ditch a discarded plastic bottle. A glance at the label: Gramox (paraquat). She tapped on the bottle with two fingers while she talked, then swiped her hand in front of her face. Marra translated: she irrigates the field before planting, allows nati-nyasi to grow, and then applies this chemical. This suppresses the weeds long enough to get the rice planted. The swipe across her face? He could not translate.

I was surprised to see paraquat in so many places. This herbicide had been used in the United States since the 1970s, but newer products were safer and more effective. Depending on the formulation, paraquat was one of the most acutely toxic pesticides. Applied correctly, it burned the tops of weeds that appeared above the soil surface. Applied incorrectly, it could kill your crop. Breathing or ingesting it could burn your lungs and maybe kill you.

Paraquat has been a health problem in countries where poor working conditions, improperly maintained equipment, extreme temperatures, low literacy, and poverty contribute to unsafe use. The manufacturer considers paraquat safe if used according to label directions for mixing and handling. In some countries, this safety is generally assured. Yet thousands of people around the world have died after ingesting or experiencing dermal exposure to paraquat. Thousands have suffered acute and chronic effects from exposure during improper mixing or application. In parts of Africa, India, Sri Lanka, and Southeast Asia, it is commonly associated with suicide.[10]

We got back into the Land Rover and drove off. Marra and Mr. V grooved to the beat of the same silky tenor sax that had entranced us for a couple weeks. We buzzed down a wide-open highway, Kilimanjaro peering above clouds on one side, expanses of rice on the other. On one side a landscape of stunning beauty. On the other the agro-dealers, the farms, the paraquat. Millions of tourists revered this iconic mountain. In its shadow, agricultural suppliers delivered Chinese formulations of pesticides to dealers without providing training in their use. The farmers were desperate to control nutsedge. They didn't know—or simply ignored—the risks to themselves and their family. Something about weeds made it possible—maybe logical—to disregard potential harm. Especially with the snow cap above the clouds, the weeds, and those who worked the fields, disappeared out of focus and got lost in the background.

Putting Trust in a Tuber

It didn't matter how much paraquat those farmers applied to their rice fields. Paraquat might suppress nutsedges but would not get rid of them. Farmers would get a satisfying crispy-dead bunch of leaves. But the rest of the plant would survive. In a few days, the tubers would send up new shoots and leaves to make new tubers and more leaves.

The nutsedge bean-sized tuber plays the role that seeds play for annual weeds. Tubers lie dormant in the soil during periods of excessive heat, drought, cold, or other conditions unfavorable for growth. Tubers move with soil erosion, flooding, and transport by animals and machinery. Anytime there's moisture and warm temperatures, they sprout new shoots from small buds on the tuber's scaly surface. New shoots push above the soil surface to begin a new plant. If the stem is cut off by tillage, hoeing, or a herbicide treatment, the tuber simply pushes out a shoot from another bud.

Tubers are connected to a bulblike structure at the base of the mother plant by way of a scaly underground root (rhizome). In loose fertile soils these rhizomes can extend as far as than 60 centimeters (twenty-four inches). The end of the rhizome swells to form a new tuber or a new basal bulb where new shoots and leaves originate. In one growing season, a single tuber can produce 500 new basal bulbs and over 1700 new tubers in an area of one square meter (1.2 square yards). At this rate, more than 6 million new basal bulbs and 2.5 metric tons (2.8 tons) of tubers could be produced in one year from a hectare (2.47 acres) of land.[11]

In addition to tubers, nutsedge produces seeds. Likely those Tanzanian rice paddies were full of them. Fields with high nutsedge infestations can have 600 million seeds per hectare (240 million per acre).[12] They germinate over a long period to produce a new cohort of seedlings soon after the last one was killed by paraquat. And nutsedge seeds are small; a thousand can fit in a teaspoon with room left over. They move easily, sticking to clothing, fur, machinery, tires, and mud that slides long distances.

Nutsedge seeds contain the genetic variation that is key to their wide-scale adaptation. The flowers are self-incompatible—they are obligated to cross-pollinate with flowers of another nutsedge plant. They do so readily: a single flower stem can produce more than 6,000 viable seeds. There can be more than half-a-million flowers in a hectare (more than 200,000 per acre).[13] Cross pollination produces seeds with varying combinations of genes. The seeds grow into plants that vary in hundreds of traits, including height, leaf length, flowering time, heat tolerance, drought tolerance, as well as tuber size and number. There is enough genetic variation to generate genotypes that survive in new environments. These novel gene combinations are the source of variation for agrestal selection in response to mechanical, chemical, and biological control measures.[14]

A plant with hearty tubers, high seed production, and variable genetics was poised to become a weed. People, animals, and flowing water dispersed tubers and seeds that withstand environmental extremes and can regenerate new populations if leaves and shoots are removed. All nutsedge needed was for humans to clear the soil (with tools that spread the seeds and tubers) and make enough sunlight and water available. There was no stopping it after that.

Words Fail

As we approached the town of Iringa, in the southern highlands, Marra said we were nearing the place where the Maji Maji had waged a rebellion in the early 1900s, the beginnings of Tanzania's nationalist movement. Iringa was a pleasant town with remains of German colonial architecture and a clock tower stuck at 6:14. The regional ministry of agriculture had arranged for farmers to meet us at a local cafe. Goats bleated in the background and chickens clucked across the room. The farmers—a mix of men and women—were gracious and businesslike. By now, the format seemed routine: I supplied enough warm Coke and Fanta to keep the conversation flowing, and Marra stood in the back to fill language gaps in both directions.

The local extension agent made a courteous introduction and explained that in this region, women harvested vegetable crops while men harvested the grains. After my greeting in Kiswahili, I explained my desire to learn from their knowledge about weeds. There were the usual murmurs, nods, and quick translations. I was assured by a gaunt man near the front that they did indeed have many weeds, and only rich people could afford to pay laborers to keep a field clear of them by harvest time. A woman in a black dress and veil near the back added that women did not have the same access to transportation that was available to men; women were challenged in taking their harvest to market. The woman next to her—they could have been twins—said married women farmed land belonging to the husband's family, and the family could take the land away if the husband abandoned her and the kids, or if he died. I had lost control of the discussion about weeds. Then a man near the front, in a shiny black suit and pointed shoes, explained that men and boys applied pesticides and chemical fertilizers on women's farms and helped apply water for irrigation.

That was my opening to get back to the subject. I asked if any of them used methods other than hoeing to control weeds. There were murmurs but no one seemed to know how to respond. Finally, the man from the ministry spoke up. He thanked me for visiting and for rousing their attention about weeds, and helping them to think about the need for good weeding because without it their crops were sure to fail, and they trusted God that new safe ways were being invented at far-off universities to help them grow their crops and manage weeds in an excellent manner but until that time they would surely continue to work the fields in ways that had always been successful for generations from their ancestors to now and so thank you for coming and sharing with us.

I couldn't wait to ask Marra what that was all about. He explained it something like this: there is no word in the local language for pesticide, so they use *dawa*, the Swahili word for medicine. They do not distinguish an insecticide from a fungicide from a herbicide. They are all pesticides, medicine. If the tomato crop has a disease, they apply dawa to heal it. In some dialects, the word for medicine is the same as the word for poison. If insects are eating the tomato plants, they use the dawa to kill them. Marra took a breath. The women in the back, with the black veils. They were mourning their cousin. Two weeks ago. Pesticide poisoning. It is the same color as Coca-Cola drink. When farmers finished spraying, they put the leftovers in bottles. Marra had heard people talking. Some said the cousin was sick; maybe he needed medicine. Some said it leaked on to his skin from the knapsack sprayer. Some said it was not an accident. Marra went on: "They thought you were going to talk about this dawa, this weed killer; to tell them to spray it on their fields." He paused to let this sink in. "I knew they were wrong. So, I didn't warn you. I'm sorry."

Hoes and Goals

The burden carried by women in small scale crop tending has been a fact of smallholder life around the world since the invention of agriculture. The belief that the first crop farmers were women has romanticized the beginnings of agriculture as simple, clever, and quaintly domestic. The anthropologist Hermann Baumann described agricultural beginnings as "hoe culture." He attributed "the most ancient method of soil cultivation . . . to the female sex. . . . Woman . . . noticed shoots growing from pieces of root and . . . with womanly economy, she artificially induced the repetition of this natural process, and planted slips of roots . . . and we see the first form of agriculture." This led nomadic hunter-gatherers to "settle down, and their wandering mode of life [became] bound to the soil."[15]

Like other descriptions of early crop culture, Baumann didn't explain where the hoe comes in. After planting the crops, somebody had to stick around and hoe out the weeds so there would be something to harvest. Men spent many days away, sharpening arrows, and sipping palm wine on their jaunty—and frequently unsuccessful—hunts. Women and children stayed behind to hoe and pull the weeds. Frustrated by unreliable supplies of food, women found ways to provide dependable sources of calories, even if that meant they had to pull the weeds themselves.

About twelve thousand years later, issues of women in agriculture were addressed in the initial eight United Nations Millennium Development Goals for countries that depend mainly on subsistence agriculture.[16] In 2000, targets were set around goals for women's education, gender equity, health, civic engagement, and other issues. There was no mention of what occupied women in subsistence agriculture. Nevertheless, many NGOs won contracts to implement projects to advance a model of "development" based on adoption of technological solutions and deployment of commercial products.[17]

I found myself inadvertently tangled in the web of these Development Goals when I crossed Tanzania to locate sites for field trials. The NGO I volunteered for needed the trials to demonstrate agricultural products, like herbicides. Getting products into the hands of farmers would help the NGO reach their Development Metrics.

A new set of 17 Sustainable Development Goals, with 169 targets and numerous metrics, were launched by the UN in 2016 (subsequently labeled the 2030 Agenda). The goals are based on a belief that "growth" will reduce poverty and inequality when smallholders produce and consume their way to environmental sustainability. It will be done with technology implemented through private enterprise and private-public partnerships.[18] One technology to be implemented to ease the labor demands for women and increase crop yields is GMO crop seeds and their accompanying herbicides.[19]

At the time, GMO technology was not legal in most African countries. Also, smallholder farmers were accustomed to saving seeds from year to year, a practice that is not compatible with hybrid patent-protected GMO seeds. A follow-up study in sub-Saharan Africa reported that pesticide inputs did increase crop yields and income. Yet they also increased health expenditures and negative impacts on the surrounding environment. The net result was a decline in productivity and well-being.[20]

Coming of Age in America

No weed could be ranked among the world's worst without going through the grinder of selection and dispersal in American agriculture. Spaniards carried weedy nutsedge along with edible chufa into the West Indies around 1500, not knowing the species had reached the western hemisphere long before Columbus.[21] Floating nutsedge seeds and tubers likely crossed from Africa to the Americas with sea currents on drifting masses of soil and vegetation. Indigenous people had used tubers—chufas—as a food for more than 11,500 years. In fact, genetic evidence suggests that nutsedge was introduced in the western hemisphere many times from various parts of the world.[22]

Even as European agriculture took over the continent, yellow nutsedge wasn't initially recognized as a weed. An early observation of nutsedge in North America was written by the Dutch-born American explorer Bernard Romans. He traveled across Florida just before the American Revolution to describe the agriculture among remnants of the Chickasaw Nation. He described a useful grasslike plant, on which horses fed. It was called "nutt grass" after the tuber and leaf: "This, when once it takes in the ground . . . makes a very good pasture. . . . but it must be well fenced against hogs, which being very fond of the nuts would root all up in a short time."[23]

The southern agricultural press of the 1860s published many articles about growing nutsedge, for the tubers, to feed hogs and chickens.[24] Asa Gray's 1862 manual of plants in the northern United States didn't mention yellow nutsedge, suggesting it wasn't common before the Civil War.[25] Even L.H. Dewey's 1895 book on weeds didn't include yellow nutsedge in a list of a hundred troublesome plants.[26] The earliest description I've found of yellow nutsedge as a weed is Robinson and Fernald's 1908 update of *Gray's Manual*. They gave it only tepid condemnation as "sometimes becoming a pest in cultivated grounds."[27]

In North America, the introduction of tractors, followed by the invention of herbicides, finally make yellow nutsedge a respectably despised weed. Tractor-driven disks, rakes, and harrows dragged the underground nutsedge rhizomes along with tubers to parts of the fields that weren't already infested.[28] When farmers started using herbicides, they quickly learned something about nutsedge biology. The first group of herbicides (2,4-D) killed broadleaf weeds, but not grass weeds; the second group of herbicides (dinitroanalines) killed seedling grasses, but not broadleaf weeds. Neither type of herbicide controlled nutsedge. That's when it became clear that nutsedge was neither a broadleaf nor a grass species. It was something different, something worse, something that didn't respond to the miraculous chemicals that were supposed to make weed control easy. Instead, the herbicides killed most of the other weeds. That gave nutsedge access to the space, light, nutrients, and water that it needed to fill fields with its tubers and seeds.

When I came along as an academic weeds guy, I kept a wary eye on yellow nutsedge. I hoped I didn't have to deal with it and cursed it when it invaded my field experiments. I've confronted it in major agronomic crops like corn, where conventional farmers usually spray it with herbicides, and organic farmers try to beat it back with mechanical cultivation. I've seen it poke through thick layers of mulch and pierce thin layers of plastic that suppressed other species. In vegetables and other short-statured crops that produce little shade, nutsedge grows wildly, soaking in the sunlight, water, and nutrients intended for the crop plants. I responded to one case of a tomato grower who thought it was a grass, treated it like a grass, and returned two weeks later to find the grasses dead while thousands of dollars in tomato transplants struggled beneath nutsedge as thick as a hay field.

Life Would Be Impossible

Farmers long recognized yellow nutsedge as a weed that thrives in crop fields and escapes control measures. But it didn't make sense that such a short plant—rarely over 30 cm (12 inches) high—could suppress the growth of tall crops like corn, rice, and soybeans. Nutsedge is efficient. It doesn't produce tall stems that might be vulnerable to damage by pests and predators (and farmers). Yet it can stunt surrounding crops and turn them yellow.

Nutsedge does this through allelopathy (uh-LEEL-o-pathy), the release of chemicals that inhibit the growth of neighboring plants. Suppressing plants with its own herbicide (allelochemicals) allows nutsedge to steal more light and nutrients. Even if the aboveground parts are removed, nutsedge roots and tubers can still release allelochemicals. Those chemicals linger in the soil and continue to suppress the growth of crops.[29]

Nutsedge also makes chemicals that inhibit the sprouting of buds on nutsedge tubers. It might seem counterintuitive. But tubers have a dozen or so buds (like "eyes" on a potato), and there can be thousands of tubers in a field. If all the buds on all those tubers sprouted at once, they'd exhaust soil nutrients before any reached full size. By inhibiting growth of some of the buds, nutsedge allows a few buds to grow into large, healthy plants. This saves back the next generation of dormant buds to grow later, which insures its future reproductive potential.[30]

Allelopathy is just one way nutsedge has used remarkable chemistry to become a successful weed—and a valued crop. The plant is essentially a biochemical factory. It cranks out all sorts of chemicals that are not essential for plant growth; they're often called "secondary" chemicals, byproducts of ordinary metabolism. Secondary chemicals blur the line between weeds and crops. They include oils, saponins, terpenoids, and alkaloids, that deter predators and attract pollinators.[31] Secondary chemicals are also the source of flavors, fragrances, colors, oils, soaps, and pharmaceuticals. They're the reason for many relationships established between plants and humans, who provide protection and dispersal to new environments.

While nutsedge churns out chemicals to make it a more despised weed, the edible chufa type uses the miracle of chemistry to be a more desirable crop. Chufa plants put antioxidants with anticancer properties into tubers along with sugars, starch, and fat.[32] Chemicals in chufa extracts have also been used as a perfume, soap, diuretic, stimulant, as well as a treatment for indigestion, and dysentery.[33] The tubers are a botanical remedy to improve blood circulation.[34]

Food and health properties of chufa have made it a popular crop in many countries. It's common in markets from Egypt to the Mediterranean and into West Africa. Like Bordeaux wine, "Valencian chufas" from the Valle de Nito in Spain are legally protected with a denomination of origin. They're used to make *horchata de chufa*, a frothy drink that originated with the Moors. Horchata has become trendy at high-end food boutiques and health stores in the West. A similar drink, *kunnun aya*, is made in parts of Africa.

The most stimulating research about secondary chemicals in chufa tubers comes from their long use in traditional medicine as an aphrodisiac.[35] Diets supplemented with ground-up tubers increased sexual motivation, copulatory behavior, sexual performance, serum testosterone levels, and antioxidant activity in male rats.[36] So far, no research confirms similar benefits in humans. Nonetheless, it is an ancient Arabic practice to give the gift of chufas, *hab al-zulom* (seeds of men), to new grooms.[37]

Pulling at the Root

I ended my Tanzanian assignment with a few days cooped up in a cubicle. I was to write a report that would capture what I had seen and learned. The NGO administrator asked me to make recommendations for "how to modernize weed management" in East Africa.

When Marra dropped me off at the office, he said we would not see each other again; I would stay in the city, and he had another assignment. After many days together, he had revealed little of himself. The Maasai were nomads, travelers, he said. They had to learn the language and ways of people wherever they went. His parents would not let him continue his schooling, so he became a driver, a traveler. I gave him a gift to pass along to his young daughter, and thanked him for his help as a translator, tour guide, entertainer, and safe driver, true to his word. He handed me a thin stack of papers; I might want them for my report.

The next morning, over a cup of milky sweet African tea and a samosa, I looked through them. They were mostly advertisements for agricultural products from dealers we visited. There were newsletters and newspapers from towns we had driven through.

From an unmarked folder, I pulled an anonymous four-page story on plain paper: "A Useful Herbicide Becomes Deadly Toxin to a Whole Acre of Maize Plants—A Story from Popote." That was the first town we had visited. A grainy picture of a downcast woman was captioned "Mama Julia." Below was a picture of what had once been a field of maize, the knee-high plants shriveled brown. According to the story, Mama Julia had hired laborers to hoe the field clean. But nati-nyasi soon reappeared. Mama could not afford to hire the laborers a second time. The man at the agro-dealer advised her to apply a chemical to kill all the weeds at once. It would not cost as much as the laborers. She needed the money she would save. Her husband had died the previous year. She suffered from paralysis. So she bought two bottles of the herbicide and hired a man to spray her maize field. She, herself, had not been able to see the result. The next day her son and neighbors told her what happened to the maize crop.

I had heard stories like this before. The pictures were vivid. The article went on to promote a different herbicide, which would not kill the maize. It didn't mention that it would have been useless against nati-nyasi. I didn't know what to make of it. Marra had wanted me to see this. It was stapled to an advertisement for paraquat.

I put in a call to faculty at the agriculture university. I had met them years before, and they agreed to see me the next day. After just a few weeks in the country, there was no way I could understand the dynamics that linked weeds to ruined lives. I wondered if these stories were as common as they seemed, and what I might say about them in my report. I needed help from agricultural scientists who understood the culture.

Three academic colleagues met me in an empty classroom. I described what I had seen from one end of the country to the other in just a brief time. I told of women who resented the job of weeding the family farm; unsafe storage and application of herbicides; people poisoned, and crops destroyed. It seemed like a crisis and I wanted to be sure I understood the situation correctly.

They listened politely. When I was through, the rural sociologist, an elegant woman with piercing eyes, glanced at her colleagues and back to me. She folded her hands and spoke stiffly: It's all about training. With proper training, these things would not happen. As for suicides, people will always find a way. She said if I wrote in my report that farmers should not use herbicides, I would be saying women must stay in the field and hoe weeds the way their ancestors did hundreds of years ago. Colonial powers had purposefully kept them at a subsistence level, so they would be poor, easily manipulated, and dependent. They must modernize. Yes, there will be errors. Some crops destroyed, some high pesticide residues. There will be some fish killed and some people poisoned. The same things have happened in the United States, she glared. But the farmers must have the same technology American farmers have. It is their right. To suggest otherwise, I must give them another way to manage weeds.

Returning to the cubicle to write my report, my mind wandered back to the first time I visited an African farm. I was a fresh-out-of-college Peace Corps volunteer. My friend Djas, a slender, toothy Ghanaian, took me to a hillside spotted with cocoa trees, papayas, and an occasional banana plant. I could still picture his farm, an asymmetrical field with irregularly planted maize. At the time, it seemed like chaos. Between maize plants were woody stems of cassava. Djas twisted a plump ear from a stalk of corn, then bent the plant over so the neighboring cassava plants could get more light. Where he had harvested cassava, he planted beans and sweet potatoes. Too little light reached beneath the dense canopy to allow for many weeds. There were a few, mostly amaranths and nightshade, which he harvested as vegetables. Nearby was his wife's farm, a patch of tomatoes, spinach, chard, beans, cabbage, onions, red-hot peppers, cocoa yams, and herbs all surrounded by pigeon-pea and lemongrass. The rains had come on time and there would be enough for the family and some left over to sell or trade.

Djas smiled and asked what I thought. Coming from the orderly monoculture of Midwest farm country, I didn't appreciate the beautiful complexity before me. I mused that I had never seen crops that were not planted in long, straight rows. He paused and looked around. Along with a few gawking chickens, there was an assortment of functionally diverse species that would provide more than adequate sustenance for his family. His wife's plot had another dozen or so species, adding nutrients, variety, and spice to their diet. I never forgot his response: "But why?"

That was a question I couldn't answer in my report. Straight rows of a uniform crop so you can drive a tractor through, spray the weeds, and hope you don't kill your crop or poison the well. My colleagues at the university believed in the approach preached by missionaries of Western technology. But why? African ways, as Djas had shown me, were ingenious, resourceful. He grew crops to feed the family before feeding the exchange system. He traded time, labor, and goods by reciprocity rather than profit seeking. There was dignity in his labor, his sense of community. I couldn't help wondering if dependence on imported pesticides, GMO seeds, and steel machines that run on diesel and oil wasn't just another kind of colonialism.

I submitted my report and made a formal presentation to the project leaders. Mr. V was there, all smiles in the front row. After thanking him and everyone for their kindness and generosity, I made clear my intention to be objective and not critical. I recounted the things we saw and heard about pesticide safety, storage, application, and training. I described the practice of selling toxins to agro-dealers and farmers without training them in how to store and apply the chemicals. I cited studies showing that where paraquat was banned or restricted, deaths from suicides had dropped.[38] I found that weed-killing chemicals were not well understood by farmers or NGOs. I hoped the attention being given to women would not overlook their dilemma—laborious hoeing or dangerous herbicides—and the consequences of both for their lives and the lives of their children.[39]

The NGO director's eyes had glazed over. Mr. V was playing with his watch. I tried one last pitch: weeds are not like insects. Herbicides cannot be used the way farmers use insecticides. I offered to return, to conduct training for farmers and dealers on how to use herbicides properly. I offered to analyze the data from field trials. The NGO administrators assured me: they would write to me. Mr. V would send data from the demonstration trials. The director would seek my advice. We all smiled. We shook hands. We promised to keep in touch. And I knew I would never hear from them again.

By the time I got to the airport, the incubation period for whatever was in the cold egg sandwiches served with tea had reached a critical stage. I changed my seat over the wing for one right next to the bathroom during the thirteen-hour flight. A fitting end to the journey, perhaps. Over the next few weeks and months, I sent messages to Mr. V., the project supervisor, and everyone I had met. Only Marra responded, asking me to recommend him to anyone going on safari. A couple of months later, I learned that the farmer-to-farmer program had ended when the consultancy disbanded. A new NGO had taken over the contract with USAID. The objectives had changed. There would be no weed trials and no safety training. Quite possibly, they had known this all along. A tour by an expert from an American university was a metric to check off and put into the final report.

Alternative to the Alternatives

I've spent most of my professional life looking for alternative ways to manage weeds, other than hoes and herbicides. One option that has occupied me, and is relevant to yellow nutsedge, is biological control. Unfortunately, the insects that feed on nutsedge are unsuitable controls, as they also feed on crop plants. The few insects that are host-specific either don't do enough damage or have been difficult to sustain. Moths, stem borers, and leaf miners have been tested, but none cause enough damage to suppress nutsedge in a useful way.[40]

I tinkered for a while with fungal pathogens as biocontrol agents. There was hopeful research on a rust fungus (*Puccinia canaliculata*) that attacked nutsedge. Natural infestations of this fungus occur too late in the season to cause much damage. But if the fungus was introduced early in the season, it could infect and kill nutsedge plants. It seemed to work in the field. Sometimes. But the host range for the fungus and variability in nutsedge susceptibility turned out to be greater than first thought, so it is not used.[41]

Biocontrol organisms that attack insect pests and plant pathogens have advanced greatly in the last few years. But none that attack weeds. This isn't due to lack of interest on the part of weed researchers. The biological and economic barriers are high. Billions of dollars have been invested in herbicide development and marketing. Very little, by comparison, has been invested in biocontrol of weeds. There's little profit potential in a biological control organism that causes sustained damage and keeps a weed like nutsedge in check. And that kind of product wouldn't fit the economic model for private sector engagement in the 2030 Agenda.

Getting It Right

Another chance to work in East Africa fell into my hands a few years ago. The USAID-funded project aimed to advance the use of Integrated Pest Management (IPM) technologies in vegetable crops, where pest pressures are especially high. The call for proposals emphasized the need to reduce pesticide use and to promote gender issues, especially women's health. Finally, I thought, they were addressing key issues of concern to women, like those I had met across Tanzania. The new project would focus mostly on insects and plant diseases, but to address the gender equity issues, I included work on weeds. With stories of paraquat misuse in mind, I inserted a plan for training on safe pesticide use, especially herbicides.

During the initial project visit, I retraced some of the paths that Marra, Mr. V, and I had traveled. Women were still out in the fields slinging hoes, children resting under a tree. I visited agro-dealers and found the same Coke bottles with unlabeled brown liquid. The teen behind the counter did his best to look up from a mobile phone. The stats on poisonings and killed crops had not improved. This project would give me a chance to work with local experts to make a dent in these issues.

I chaired the initial meeting with about thirty project participants. We were to prioritize crop pests, and to agree on the metrics for participation by women. Long discussions began about insect problems and plant diseases that were new to me. I kept my mouth shut and listened. Insect-vectored viruses and pesticide resistance seemed overwhelming. The American and African experts in crop insects, diseases, viruses, and nematodes put in their bids to focus on their pests of choice and debated the order of priorities. Next, the sociologists and economists took over, expounding on ways to ensure participation by women farmers and to analyze data at the family level.

I was about to close the meeting, but I couldn't resist one question: if we're concerned about women and crop pests, shouldn't we consider weeds? Especially weeds like nutsedge that the women are out there hoeing?

I guessed this would get no traction. I just wanted the experts to respond. After a few sighs and coughs it came: for crop insects and diseases, there are many biological controls, botanical insecticides, parasitoids, and other beneficial organisms. These are all safe technologies. There is nothing similar for weeds, only hoes and herbicides. So weeds cannot be a priority for this project. Yes, there is nutsedge in the fields. Maybe all of the fields. The laborers hoe them and pull them out. That is what they do.

End of discussion. It was time for the latest directive out of Washington from the policy folks. First of all, they explained, Tanzania recently voted contrary to the wishes of the US government at a recent session in the United Nations. Future funding for its projects was uncertain. Secondly, we needed to revise the project proposal. The part about pesticide safety training had to go. It sends the wrong message. The Chinese had the pesticide market in Africa. We didn't want to be promoting that. Under the revised policy, we needed to work with American NGO private sector partners, to push American technology adoption. We needed to get our metrics in step with the administration's goals.

Edges of Agri Culture

Driving down the Mbeya road on my last visit to southern Tanzania, we pulled off at a small sign marking the Isimila Stone Age archeological site.

It's at the other end of the country from the more famous Olduvai Gorge, and can't compete with the Olduvai's 1.8-million-year-old *Australopithecus* finds. Isimila is located in a deep eroded gulley where angular, blocky sandstone pillars rise to more than thirty meters (greater than ninety feet) in height. The site is important for its collection of tools, massive hand-axes and hoes dating back 60,000 to 100,000 years.[42] Not dainty, sharp-edged metal tools, these were essentially rocks the size of large bricks.

On first glance, I wasn't impressed. Who could say these sharp-edged rocks were tools? Yet examining one after another—there are thousands in the little museum—revealed a consciousness behind the flaked edges. The flakes weren't just on any random side of the stone, they were positioned so that the other side could function as a handle. I walked deep into the steep dusty valley where the tools had been found. Toolmakers, interpreting shape, color, and texture, knew where to strike one on another to make a commonplace stone into something new. One side an edge, a blade; the other side a handle.

They were used in hunting. They were used to scratch the earth to uncover roots, corms, and tubers, sources of nourishment. They represented the human influence on the planet that began long before the idea of settled agriculture. Prior to conscious selection of plant morphologies was mindful selection and manipulation of things of the earth. What first looked to be a rock could be a tool. What first looked to be a useless plant could be food for today, and one to seek again tomorrow. Continual choice of one morphology—the particular feel, direction of in the grain, or angle of a surface—made a blade and a handle. Others—the ones I stumbled over walking through the canyon—were discarded. In the same way, primitive crops and weeds, like chufa and nutsedge, were born in tandem.

Since the time humans first decided to forego the uncertain results of hunting and collecting from the wild in favor of sowing seeds, protecting seedlings, and harvesting plants, the notion of crop agriculture has been synonymous with weeds and weeding. The notion of "cultivating" a crop—preparing ground, planting seeds, and nurturing the plants through harvest—has long been conflated with that of "cultivating" weeds—using a hoe or other tool to remove them in order to give the crop a chance to grow.

As we drove away from Isimila toward the noisy city and a waiting airplane, I looked across the fields. In places like this, thousands of years

ago, early farmers and gardeners first cultivated plants that became vegetables and grains. A quirk of the English language reveals the reality that
to do so they must—in a different sense—have cultivated weeds. They
cultivated, tended, and nurtured plants cherished as crops; they pulled,
dug, and crushed plants derided as weeds. Some, strangely, were one and
the same species. The land moving across my window was flat and rocky.
Laborers, men and women, were busy turning soil, slashing grasses, hoeing weeds. Their toil wasn't so much about making the crop grow and
thrive as it was about trying to ensure that weeds did not. They worked
the fields because weeds demanded their effort. It's an effort so fundamental to the mystery of securing a harvest, that the weeds and the laborers
could fade from view. From this primeval landscape, the production of
food, fiber, flowers, and herbs, along with the destruction of weeds—
tubers, leaves, and all—appeared as opposite, yet synonymous; it was
impossible to imagine one without the other.

Marestail, *Conyza canadensis*.

5

MARESTAIL

In the early 1990s, a Saturday morning in April might find me on Main Street in Smithville, Ohio, somewhere between the Seed-N-Feed near the railroad tracks and the Hardware at the stoplight. The street cuts through the center of a farming town on the eastern edge of the US Corn Belt where gently rolling hills of Northeast Ohio meet the glacial moraine. The town and surrounding county have held each other together since 1818, now a community of about two thousand mixed grain and dairy farms.

I'd stop into the Hardware—a daughter in tow—to pick up a bag of chicken feed or a couple of fence posts. The red brick two-story, built in the 1830s, anchors the town. The front window announced sales throughout the year: canning supplies, tulip bulbs, bird seed, and such. A bell jingled above the door and we'd walk across the creaky oak to jars of seeds, bins of onion sets, and bags of seed potatoes. We lingered to overhear the chatter among local farmers. They talked of serious things—sick cows, farm auctions, commodity prices, and the weed that got away last year. Farmers might prod me, asking what new mix of weeds I was putting in the fields that year. My standard response: "It's up to you; whatever weeds you choose will be there waiting for you." Sometimes this got a laugh.

We might drop into the IGA grocery next door, a good place for ice cream or a fried fruit pie, then head out past the Seed-N-Feed. Family owned for more than 160 years, the Seed-N-Feed was a typical Midwestern full-service seed supplier, grain elevator, and warehouse for custom application of fertilizers and pesticides. It connected major agriculture corporations to the farm landscape, the conduit for the newest corn hybrids, soybean varieties, and herbicides. On springtime mornings—except for Sundays—mammoth yellow fertilizer trucks and giant insectlike highboy pesticide sprayers buzzed in and out of the dusty gravel parking lot.

Back then, Smithville flaunted its small-town ways, its history of Swiss dairies, and connections to Amish and Mennonite communities. Driving in and out of town, I watched the fields change with the seasons. I watched over the years, too, with a view to the changing weeds. I saw the steadiness of the town fading when the countryside began to change. It didn't take long. Marestail was spreading on the best agricultural soils.

Marestail had always been around along field edges and waste places, and nobody gave it much attention. It wasn't like dandelion, velvetleaf, beggarweed, or nutsedge; it had never been considered useful or attractive. It was just one of those plants that colonized disturbed ground and then faded away. The stiff stems with narrow leaves pointing all directions went unnoticed until fall, when whitish grey puffs swayed in the wind. In the puff are thousands of small seeds with a bristle, or pappus, that acts like a parachute. Wave a marestail plant in a breeze and hundreds of seeds fill the air—many more than blow from a dandelion. They fly high and settle slowly: marestail seeds could cross Ohio from Smithville to Indiana and still have a leisurely landing.[1]

Nobody looked twice at marestail when more troublesome weeds were around. It would take hundreds of marestail plants per square meter to do any harm.[2] And just a little tillage in fall or spring kept marestail away.[3] The small, mostly self-pollinated flowers would never make new and interesting variants that might attract attention.

And the name was confusing: "marestail" and "horseweed" were used for several other weeds. Canadians call it "Canada fleabane," but it's not a fleabane. The scientific name, *Conyza canadensis,* means a strong-smelling plant from Canada.[4] Some people insist on calling it *Erigeron*: "Er" signifies spring, and "Geron" an old man; an old grey-haired plant in springtime.[5] But it doesn't turn grey until after flowering in midsummer.

In the 1990s, marestail woke from its sleep and covered the farm landscape. This humble, unassuming, native plant became entangled in a web of opportunity and good intentions that drove unanticipated shifts in species, genetics, technologies, and communities. In just a couple years, when prodded for a weed prediction I had a sure response: Marestail. And nobody laughed.

Shifting Soil, Shifting Weeds

The rise of marestail began with a few gusts of wind. Actually, a lot of wind. Following recurrent drought and crop failures in the early 1930s, the American Dust Bowl blew precious soil resources of the Midwest to the East Coast. Nearly 3 million people were pushed off farms.[6] The government of Franklin Roosevelt responded to the social and ecological tragedy with a series of programs. One was a system to manage grain supplies, especially overproduction, which drove prices down and was hard on the soil. Next were programs to advance soil conservation, which instilled notions of land stewardship in generations of farmers and agricultural researchers.

Farmers' responses to Dust Bowl programs were mixed. Many turned to Edward Faulkner, whose *Plowman's Folly* proclaimed that plowing was unnecessary and harmful to soil. Faulkner, an agricultural educator in Ohio and Kentucky, poked a finger at conventional practices of the day: "The moldboard plough has been shown to be the villain of the world's agricultural drama."[7] The question, then, was how to farm without the villainous plow. Farmers had no way to plant crops into unplowed ground. And there was no way to control weeds without first plowing, harrowing, and cultivating the soil.

Faulkner's ideas helped motivate the development of no-till and minimum-tillage agriculture.[8] No-till advocates still cite Faulkner's famous line: "No one has ever advanced a scientific reason for plowing." The critical innovation that made no-till agriculture possible was chemical weed killers that could substitute for the plow. So-called "burndown" herbicides would destroy all plants before crop seeds were sown. This gave crops a clean start. Other (preemergence) herbicides were needed to kill weeds that germinated before the crop emerged from the soil.

The invention of 2,4-D and other herbicides set off a storm of government, university, and industry research to advance no-till practices. By the 1960s, the chemical approach to farming was encouraged by relatively cheap fertilizers and pesticides along with precision planters and high horsepower tractors. The expensive equipment led to specialization, monoculture, and larger farms. Banks lent money at low interest rates for annual outlays of production inputs and new machinery. Those who were slow to get on board were first on the auction block.

As a student of agronomy in the 1970s and 1980s, I had a front row seat to the experimentation that refined no-till practices. I also got a heavy dose of exhortations about its benefits. No-till reduced soil erosion by over 90 percent, especially in hilly terrain. Residues left on the surface built up a layer of organic matter. This protected the soil, improved structure, and provided a zone rich in nutrients and beneficial soil microorganisms as well as earthworms. Slice a shovel into a long-term no-till soil, and the profile resembled that of the forest floor.

Farmers adopted no-till because it promised to save time and money. For some crops, they could spray and plant in one operation, eliminating three or more trips across the field. No-till allowed farming on fields that were too steep or too wet to plow. Though crops started out slowly in a cold spring, they made up for it by midsummer, using moisture held in the soil beneath the layer of surface residue. By the mid-1990s, farmers in some regions used conservation tillage on up to 70 percent of land for certain crops.[9]

No-till farmers were encouraged by how easy it was to control weeds that were problems in the past. With no-till, small weed seeds remained buried too deep to germinate and emerge. The first couple of years after adopting no-till, many weed species declined or disappeared from their fields. But weeds change. Nobody considered the converse, new weed problems that might now appear. Plowing the land held back the process of "ecological succession," the natural pattern of plant reestablishment. When farmers stopped plowing, succession from easy-to-kill annual weeds led to hard-to-kill biennial and perennial species. It happened slowly. But by the time I started weed ecology research on the rolling hills of eastern

Ohio, in the late 1980s, this pattern of weed "species shifts" had become a major concern in no-till fields.

Seed-N-Feed

One morning in the early 1990s I stopped by the Seed-N-Feed to order supplies for field experiments. Chet, a ruddy, balding veteran, stood behind the counter, fiddling with a thick pencil. He had gone out of his way for me to find seeds of unusual crops or varieties I wanted for research. In return I identified unusual weeds his customers dropped off. This time, he pulled out a bouquet of thorns. A farmer, five years into no-till, had run into a patch of briars and poison-ivy while harvesting wheat. I was the weeds guy—did I want to go help untangle them from the machine? He tried to smile.

Novice no-tillers were surprised by the shifts in weed species when they stopped plowing and started to depend on chemicals to control weeds. Most annual weeds—including marestail—were killed easily by those burndown herbicides, like glyphosate. But when tough perennials started to get established, the love affair with no-till got complicated. Farmers turned to industry, government, and university experts for help. The only solution was more and different herbicides—full rates of glyphosate along with high rates of atrazine, alachlor, and a few others thrown in for good measure.

Into the 1990s, newspapers and farm magazines ran stories warning of environmental concerns about no-till. The improved soil structure and earthworm channels encouraged root growth and water infiltration. This also allowed chemicals to move into groundwater. Nitrogen and phosphorus were lost in runoff water. Nitrates moved into subsurface waters. So did herbicides like atrazine and alachlor. No-till had been promoted as an environmental practice. Now it was an environmental and public health menace.[10]

Next time I stopped by the Seed-N-Feed, Chet's customers had more than weeds on their minds. I walked in on a discussion among eight or ten guys in Carhartt overalls and seed company caps. They prided themselves on their good stewardship of the land, farming to protect the soil, in spite of what the media said. Many had given up the plow entirely. But killing the perennial weeds meant applying more herbicides. There was no getting around it. Newspaper articles about farm chemical pollution pointed the finger at them. They were just doing what experts, government programs, and loan officers had encouraged them to do.

Out poured a hopper-load of grievances: the farm bill, regulations, the EPA. I waited in the back, watching. Chet engaged his customers, punctuating the air with a chubby finger. The farm crisis of the 1980s was still fresh, when 20 percent of farmers had gone out of business.[11] Now, they avoided talk of foreclosures and who had just bought their neighbor's land. They grumbled about production costs, land values, commodity prices, and left-wing environmentalists nipping at their heels with more restrictions on freedom. I had gotten to know Chet and other locals but had never heard their discontent so plainly. Politics mingled with uneasiness about free trade, grain sales, chicken lots, machinery costs, crop varieties, fuel prices, the latest pesticides—along with new outbreaks of insects and weeds.

I would take up my business with Chet another time. I backed out the door and left the Seed-N-Feed. I had another group of farmers to visit, over in the next county.

Alternative Realities

My colleague, Ben Stinner, had invited me to join a meeting of farmers who were trying to organize themselves along the model of the Innovative Farmers of Iowa. Ben was a guy who could talk with anybody, gather ideas from every conversation, and find good in everyone. He asked me to just stop by, listen, and see what I thought. I followed the directions he had scratched on a scrap of paper and pulled into a meeting hall somewhere over the county line. The meeting was underway, so I slipped in near the back.

Ben introduced me as a university weed researcher, quickly adding "But it's OK, he's a good one; he's not just a herbicide guy; don't worry." That this point had to be made at the outset was a clue to something, maybe a gentle suggestion to keep my mouth shut.

The basement room was dimly lit. Folding chairs scattered in six or eight rows. Men and a few women, young and old, came from three counties around. Some were organic growers, some leaned that direction, one claimed to be "biodynamic," and others weren't sure. They seemed united by one common theme articulated by a dark-haired, rough-hewn farmer I came to know as Vince: "Get the chemicals out of our lives," he barked. "They're killing the soil, they're killing our crops, they're killing our animals, they're killing our. . . . Just get them off our farms." That was my introduction to Ohio's alternative agriculture community.

Alternative farmers had responded to the Dust Bowl in a different way. They, too, held up *Plowman's Folly* as inspiration for their response to soil erosion. They, too, cited Faulkner's famous line, "No one has ever advanced a scientific reason for plowing." But they resisted government-sponsored programs, chemical technology, and high horsepower tractors.

Alternative ideas about agriculture had always been around. Independent-minded farmers favored diverse, decentralized, chemical-free farming, counter to prevailing trends pushed by what Vince called the "university-government-industrial complex." Instead of specializing in just one or two crops, they continued to mix crop and animal agriculture, using animal wastes to stabilize soil and maintain fertility. They aimed to promote soil health through crop rotations, green manures, and other practices.

At the heart of the philosophical divide between alternative and conventional approaches to farming were weeds. Conventional farmers accepted herbicides as the foundation of their weed control program. Alternative farmers had no interest in the industrial chemical no-till model for agriculture. The whole package of technologies from hybrid corn to insecticides, herbicides, and fertilizers was expensive, dangerous, and unnatural. They would stick with crop rotation, mulches, cover crops, and a harrow with oblique rows of concave disks, as Faulkner had suggested. It was standard textbook agronomy.

For some of them, that agronomy mingled with holistic, regenerative, harmony and balance. One suggested that farming draws on vital energies that operate beyond mere chemistry and godless evolutionary forces. They shared a mix of modern and old-school ideas: deep understanding of microbial activity was linked to lunar cycles and farming in rhythm with the planet.

The discussion turned, as I feared it might, to weeds. They're "custodians of soil quality," Vince explained. "Weeds are there to tell you what's wrong with your soil. More important, the weeds are there to fix what's wrong. If you balance the soil and distribute organic matter evenly, weeds won't be a problem." I had heard this many times. To me it seemed like a step too far into botanical altruism, but heads were nodding. Vince continued: "If you spray poisons on them, you just do more damage to the soil."

After a while, a farmer I came to know as Eli described a white fluffy weed that had moved in from his neighbor's conventional field. It had to be marestail. How could he balance his soil to keep it out? He looked around, then back at me. Vince grinned a gold tooth: "Maybe our weeds guy can tell us how to rebalance Eli's soil." Everybody turned my direction. Ben shifted in his chair.

It's the most common question I get from people who don't want to use herbicides: how to balance the soil so weeds won't grow. The idea of soil balancing goes back a long time to suggestions that there's a special ratio of nutrients in the soil that's ideal for plant growth.[12] Most intriguing to me, farmers believed soil balancing controlled weeds.

Adjusting soil nutrients to meet the needs of crop growth made sense to me. It was the "so weeds won't grow" part that I couldn't square. Research and experience over hundreds of years showed that certain weeds were favored or discouraged by certain soil conditions. But no biological mechanism led a particular ratio of soil nutrients to favor crops yet suppress weeds. Weed plants enjoy the same soil nutrient conditions that crop plants enjoy. They had been selected and domesticated to thrive in the very same soil. But these farmers were out on the land. They saw things that I didn't see. And they believed soil balancing worked within management systems with cover crops and crop rotations.

I looked back at Eli. All I could do was reaffirm the good agronomic practices he already used. I said that marestail seeds could float high in the air and move over a hundred miles. If a marestail seed fell on your field, there was a chance it would germinate in spring or fall. It liked undisturbed sites that weren't too wet. That was the most accurate if unimaginative thing I could say. I wasn't sure if it satisfied or simply confirmed that I was a reductionist researcher out of touch with the finer points of soil balancing.

The meeting broke up and Ben asked to join me for the ride back to town. We turned onto a bumpy road and let the sound of stones hitting the fenders fill the silence. Back on pavement, he eased into it: "Look, maybe some of the things sounded a little absurd. They think weeds guys are all about herbicides." Ben smiled broadly; he knew some people at the university believed this as well.

Ben continued: "There's absurdity in conventional agriculture, too. Applying pesticides and not really knowing where they're going or their true environmental and health effects over time. Alternative farmers focus on different things. They're trying to close nutrient cycles and reduce purchased inputs like fertilizer and chemicals. Planting by the phase of the moon we might think is baloney, but it's an ancient practice, and harmless. Does it make more sense to pump oil and pesticides into a system that force-feeds animals and makes high-fructose corn syrup?"

I let that sink in. We turned down a narrow road, and finally onto County Road 501. Ben reasoned that one approach isn't more rational than the other. He agreed that conventional and alternative farming groups both aimed for good land stewardship. Yet they valued land and labor in different ways. "Different approaches to farming," he mused, "based on different world views, were bound to have different effects on the environment and the community. And at some level, it's all about weeds."

Life on the Edge

If there was a connection between soil conditions and marestail, nobody knew much about it. Marestail defied investigation. And why bother? It wasn't on the list of species to identify in my first college course about weeds. The standard text said it was a weed of "pastures, roadsides, and wastelands."[13] Marestail was said to lack "a distinctive personality. It is so common that it is often unnoticed. It has never established itself as a major weed problem and exists around the fringes of cropped fields, in pastures, and neglected areas."[14] In a 1990 survey, Ohio farmers ranked marestail between "none" and "slight" in importance.[15]

When no-till farmers first saw marestail on field edges they had little cause for concern, if they even recognized it. The tiny, indistinct seedlings seemed unlikely to cause any harm. In midsummer, a straight stem arose with narrow, pointed leaves. Haphazard branching at the top made a waving wand of bristly seeds by summer's end. As no-till became more common on the landscape, more bristly wands appeared. Crop residues left on the soil surface gave marestail seedlings protection. Fertilizers applied in spring and fall fed marestail during peak growth periods. Marestail genotypes that survived best in no-till fields increased in the population, a classic case of agrestal selection. But still, herbicides like glyphosate killed most of the marestail before it became a big problem. With a few basic agronomic practices—some tillage and crop rotation—it might have escaped into obscurity.

Threshold of a Revolution

During graduate school in the early 1980s, I studied among students who worshiped at the altar of plant physiology. We repeated our devotions—the chemical steps involved in photosynthesis, respiration, and other critical plant processes. Most of us weren't biochemists, but we memorized structures of chemicals that react to turn simple molecules into more complex molecules, or vice versa. To pass the exams we had to recall the sequence of chemicals, enzymes, cofactors, and numbers of ATP. It was like memorizing how parts of a car engine moved together, and luckily you didn't need to understand it in depth in order to drive.

After a while, we started to hear unfamiliar language at seminars and lectures. Speakers talked about endonucleases that cut DNA at specific places. They worked with plasmids, small circles of DNA that replicate independently. They studied regulatory sequences that controlled the action of a gene. A new set of abbreviations appeared—PCR (polymerase chain reaction) to make millions of copies of DNA strands. What passed for data was no longer numerical precision and statistical analyses we had worked so hard to learn, but blurry blotches on radiographs whose significance was supposed to be obvious.

The discoveries and new vocabulary of biotechnology moved fast. I hoped to finish up and get a job, where I wouldn't have to know that stuff. I discussed it with other overwhelmed grad students. We returned to the office after lectures and tried to figure out what we had just heard.

Where was this all going? Who cares if you can find the amino acid sequence in a long strand of DNA? What are you going to do with that?

A plant physiology student tried to explain: "A gene that codes for antibiotic resistance was moved from one strain of bacteria into another ten years ago." If we didn't know this, we were already out of date. Now researchers were trying to insert an antibiotic resistance gene into tobacco.[16] According to Mr. Wise Guy, "If you could move that gene into tobacco, you could put any gene into any plant. You could find a gene that codes for insect-deterring chemicals and put it into crops; farmers wouldn't have to use insecticides. Heck, you could insert a gene that makes crops resistant to herbicides. Maybe a herbicide like Roundup. You could spray the whole field, crops and all. The crops would live and only your stupid weeds would die."

Guffaws were muted. Skeptical. Maybe someday. Not during our career. We knew it was possible in some distant future. We all understood how glyphosate (Roundup) killed virtually all plants by inhibiting a key enzyme in shikimic acid metabolism—we had memorized that one. If somebody inserted a gene that changed key steps in that process, glyphosate would no longer kill the crop plants, but could still kill the weeds.

We couldn't fathom that we were standing at the threshold of a revolution in biology. Behind us were efforts of hundreds of scientists, starting way before Darwin, to Mendel, to Watson and Crick, to McClintock and all the rest. Here, now, was the fruition of humanity's dream to crack the mystery of the genetic instructions for plant development, functioning, growth, and reproduction. Incredible new power would be wrought from a command of the intricate workings of protein synthesis and regulation. Tinkering with genes to increase photosynthesis could make more grain that would feed a hungry world. Corn and rice plants could fix nitrogen and provide their own fertilizer. Grain crops could be perennial, so we wouldn't have to plant new seeds every year. There was endless potential to do great things.

We could not situate ourselves along the grand arc of human strivings to uncover mysteries so fundamental, a yearning at the heart of human existence. Untangling the workings of the atom, exploring the universe, cracking the genetic code: humans could not have survived without scratching such vital urges. Never mind that wisdom about what to do with all the things revealed by this scratching had not proven so indispensable. Hardest for us to accept, was that now, with humanity poised to finally pluck this shiny fruit from the tree of knowledge, how would the wise and powerful first make use of it? To kill weeds?

Prelude

The chemical weed-killing herbicide that became famous for its attachment to genetically modified (GMO) crops is glyphosate, the active ingredient in Roundup.[17] Glyphosate was first evaluated by the Monsanto Company in greenhouse trials in 1970. It had been invented twenty years earlier by a small Swiss pharmaceutical but was put on the shelf for lack of interest. Monsanto was looking for chemicals to develop new water softening agents, when a couple of them showed mild toxicity to plants. A little chemical fiddling yielded a white, odorless salt. Plants absorbed it slowly and died a week or more later. A maybe true story that went around among weeds guys was that test plants were sprayed in Monsanto's greenhouse on Mondays and evaluated for symptoms on Fridays. Unharmed plants were to be thrown out. As the story goes, the first glyphosate-treated plants showed no symptoms on Friday. But somebody was in a hurry or just lazy, so the plants sat over the weekend. The next week, dead plants revealed glyphosate's efficacy and characteristic slow activity, only because somebody had slacked off.

Glyphosate works by blocking an enzyme in a critical chemical process. Block the enzyme, and the chemicals can't link up; the process stops making the end products. The enzyme that glyphosate blocks goes by the letters EPSPS.[18] It's a key enzyme in a process that starts with shikimic acid and ends with vital amino acids. Amino acids are the building blocks of proteins. Thus, glyphosate blocks EPSPS during construction of proteins that plants need for normal growth. Without those proteins, plants die. Since EPSPS is found throughout the plant kingdom, glyphosate has the potential to kill all green plants.

When glyphosate first became available to farmers as "Roundup," practical uses were limited. Unlike other herbicides farmers were familiar with, glyphosate was nonselective (i.e., no plant was safe), it took a long time to kill target plants, and it didn't stay active in the soil to kill germinating seedlings. Compared to other herbicides, glyphosate report-edly had low toxicity, low persistence, and little potential to leach into groundwater.[19] It could be sprayed on no-till crop fields as a burndown herbicide to clear the ground of weeds before planting. Since it killed all leafy plants, there was no way to spray it over top of growing crop plants without it also killing the crop. And it was expensive.

When perennial weeds became a big problem in no-till fields, glyphosate was the only herbicide that killed them. Farmers had little choice, so they sprayed glyphosate in springtime to burn down weeds that emerged before planting. They sprayed it again after harvest to kill weeds that emerged late. It wasn't quite that easy, but the result was that farmers commonly applied glyphosate twice a year. This approach was also effective against winter annual weeds and early emerging spring annuals. Among them was marestail.[20]

The Revolution Shall Begin

Around 1994, a technical rep from Monsanto called to ask if I had room in my field plot experiments for four short rows of soybeans. He stopped by a few days later with two envelopes. Seeds in an envelope labeled "R" were to be planted in two middle rows; seeds from the unmarked envelope in two outside rows. When the plants were about twenty-five centimeters (ten inches) tall, I was to spray Roundup over the whole plot. He would stop by to check it out. In return, he left a couple of jugs of herbicide with the farm manager.

Monsanto and other companies had been working for years to alter plant genetics. With new genetic engineering methods, they invented GMO soybean plants that could tolerate applications of herbicides that would otherwise have killed them. They did this by tinkering with the EPSPS enzyme in soybean leaves, modifying the "target site" where glyphosate blocked EPSPS. The soybean's modified EPSPS enzyme would no longer be blocked by glyphosate. The crop plants would continue to crank out all the amino acids they needed for normal growth. The GMO crop would grow in spite of applied glyphosate, and all the weeds would die.

It was an elegant work of biotechnology. To change the soybean EPSPS enzyme so that it could tolerate glyphosate, they had to change soybean's genetic code, the stretch of DNA that makes EPSPS. It was like finding and changing the street number of one person in a random list of names and addresses of 20 million people. Researchers used a naturally occurring strain of bacteria (*Agrobacterium*) whose EPSPS enzyme tolerated glyphosate. They isolated the bacteria's gene (the new street number) for EPSPS and put it into soybean DNA. This tricked the soybean plant into making the glyphosate-tolerant (bacterial) form of EPSPS. It wasn't a simple cut and paste operation. They had to link the bacteria gene to something called a transit peptide, sort of like a shuttle bus. Then they added a gene promoter along with a sequence of several other genetic elements. The resulting glyphosate-tolerant GMO soybean seeds were first field tested in June 1990.[21] Four years later, GMO soybean plants were growing in two rows of my research trials.

I went out to the field plots a week after the treatment was applied. It was late afternoon in mid-July. Something in the still air said the season had already turned. I pictured in my mind what the field would look like. I had killed plenty of plants with Roundup before. I didn't expect anything especially remarkable.

I halted before a crisp, brown, stark dead patch of scorched drooping leaves—soybean and weed cadavers—and two gloriously green rows of glyphosate-tolerant GMO soybeans that sparkled in the sun. I stopped staring long enough to snap a picture and walk around the plot. The dead plants were just so incredibly dead. No sign of viable buds or green tissue in the stems. The middle two rows all smiles above the rubble.

I drove off and let it sink in. I took the long way home, past the Hardware. The window announced, "Crocks & misc. on SALE stop in & check it out." I eased over the railroad tracks and down past the Seed-N-Feed. A flag flapped above the tallest silo. The place seemed quiet for a change.

I brooded all evening over those two glimmering rows of soybeans. I remembered how GMO growth hormones had changed the local dairy business. If your neighbor used the new hormone, they'd get more milk from their cows and make more money. If you didn't do the same, they'd buy you out. When milk prices fell, the conglomerates moved in. Farmers became serfs on their own farms. I wondered if GMO soybeans would do the same to grain farmers from Smithville to Kansas.

One thing was certain: life as a weeds guy had changed. For years, advances had been made in other agronomic practices. But weeds had kept farmers frustrated, making weed control one of the most expensive and essential elements of crop production. Now, farmers would think all their weed problems were solved. So would my colleagues, my supervisor, the whole agriculture college, and funding agencies. I was approaching the middle of my career and whatever value my knowledge of weeds might have had, it just seemed so incredibly dead.

Like many of my colleagues, I had done some herbicide field testing along the way. It helped pay for research on nonchemical weed management. And it helped me stay in touch with conventional no-till farmers. Now, I needed to consider what that type of work meant for the wider agricultural landscape, including weeds and people. Monsanto made it easy for me to stop doing more herbicide tests. They required a confidentiality agreement for research that used their products. My field plots were always open to anyone who wanted to see them. I wasn't alone in refusing to sign.

Materials and Methods

In 1996, "Roundup Ready" soybeans launched the largest agricultural field experiment ever conducted across the American heartland. It's still going on. The objective of the experiment, for industry, was to determine whether fields of soybeans modified with a bacterial gene that allowed them to tolerate glyphosate could be as productive as conventional soybeans and open vast lucrative markets for GMO crops.

Farmers adopted GMO soybeans faster than anybody imagined possible. The pump had been primed by inescapable marketing. Nearly everybody at the Seed-N-Feed wore a "Roundup Ready" cap. And the technology arrived at a time of relative optimism. Commodity prices were on the rise in anticipation of new markets for American grains. Some European and Asian buyers refused to accept GMO crops; this held up the price of conventionally grown grain. Midwestern farmers were sure that objections to GMO crops would be overcome once the productivity and environmental benefits were understood. Genetic engineering was about to solve a whole lot of farming problems.[22]

The marvel of genetic modification allowed farmers to apply just one herbicide—Roundup—right over top of their crops. Every weed would die, and the crop plants would live to bask in the sunshine. They'd seen it on the TV ads, and in live demonstration plots. Now they'd be able to use no-till on more of their land, applying Roundup first as a burndown herbicide, and then again over top of the growing crop—three or more times if needed. It was that easy.

More GMO miracles were on the way. New "Bt" corn hybrids had a gene that produced an insecticidal toxin. Instead of spraying insecticides to kill corn borers, farmers could simply grow the GMO-Bt corn. And the Bt corn was also glyphosate tolerant. Farmers were sure the new technology would increase productivity and profits. It would reduce production costs, soil erosion, and pesticide use. That's what the company reps, agriculture experts, radio and TV ads, and magazines said. Best of all, it poked a finger in the eyes of liberal tree-huggers who had been carping about the environmental harm caused by no-till agriculture.[23]

Down at the Seed-N-Feed there was more cause for optimism. Rumors said a new government farm bill was in the works. According to Chet, it had a provision called "Freedom to Farm." Restrictions on all-out production would be reduced or eliminated. The joke—and hope—was that the EPA would be eliminated, too, and bring back the pesticides that had been taken from them. Chet smirked and shook his head: freedom to farm means freedom to produce more and profit more. He had plans for another grain silo to handle the abundance of GMO grain.

Farmers jumped into the new technology. This was the answer to their prayers. With larger acreage and the need for off-farm jobs to stay afloat, they were under incredible time pressure. Some, along with their spouses, worked two jobs in addition to farming. They were constantly riding a line between massive debt and foreclosure. Glyphosate-tolerant crops offered a small relief valve because it made weed control a lot easier.

Moreover, farmers felt they had little choice. If their neighbors planted GMO seeds and produced more grain in less time with fewer weeds, they had to follow. They also had to go along with the conditions Monsanto imposed. Farmers who purchased herbicide tolerant seeds had to sign restrictive agreements and pay a "technology fee." They had to buy Monsanto's brand of glyphosate, not cheaper versions sold by competitors. They agreed to plant all the seeds they purchased and could not resell them to other farmers. They also agreed to sell back to the grain company all the grain they produced. None of it could be saved for planting the next year. Monsanto lawyers came down hard in highly publicized cases of a few who tried to cheat.[24]

Orders rolled in for GMO glyphosate-tolerant seeds. Among my weed science colleagues, this was a good time to advise farmers about herbicide *resistance*: heritable genetic changes that allow weeds to survive a dose of herbicide that would otherwise kill them. "Resistance management" practices could avoid that possibility before millions of acres were sprayed yearly with glyphosate. But industry pushed back. The idea that resistance was even possible might scare customers. The way to avoid resistance was to reduce selection pressure—don't use glyphosate all the time. But no company would release a new product and then tell farmers not to use it. Plus, recent research had suggested that "genetic and biochemical constraints . . . preclude the evolution of glyphosate resistance in weed populations."[25]

And besides, glyphosate had been used as a burndown herbicide for twenty years, yet no weeds were resistant to it. Another paper was published in 1997, after glyphosate had been used for one season on GMO soybeans without a resistant weed outbreak. It concluded that "the probability of evolution of glyphosate resistance seems low." It gave the impression that there was no need to change practices to avoid resistance. Those who suggested otherwise didn't understand the complicated underlying biology.[26] In other words, concern about resistance was futile.

To those of us who conducted field-scale weed studies, the massive glyphosate-tolerant crop experiment was a research bonanza. It was a chance to find rare mutations that allow a weed to resist a herbicide destined to be used more widely than any other. We knew mutations were out there somewhere in the vast populations of genetically variable weeds, even if they were only one in several million. There was only one way to find those mutations: spray glyphosate on as many acres of weeds as possible. And now, farmers were going to do this for us.

The numbers were in our favor. There were about 26 million hectares (64 million acres) of soybeans and 32 million hectares (80 million acres) of corn waiting for herbicide tolerant technology to be deployed. If farmers planted GMO seeds on only 80 percent of those fields, glyphosate might be sprayed one or more times on about 47 million hectares (115 million acres) of land every year. Every one of those hectares likely had 50 million or more seeds (20 million per acre) of any one weed species.

That gave us at least 6 billion opportunities per year for a mutation that would confer resistance. With those odds, and agrestal selection pressure from glyphosate sprayed every year, a mutation for resistance would certainly show its face. When it did, and if it survived to produce mutant seeds, the resistant genotype would spread like, well, a weed.

I speculated along with other weeds guys about which weed would be the first to evolve resistance. That's about as exciting as conversations got at our professional meetings. The discovery of a glyphosate-resistant weed population would challenge industry's foot-dragging on resistance management. We knew, of course, that when resistant weeds appeared, industry could sell additional products to kill them. This clarified what the foot-dragging was all about.

Our resistance guessing game had focused on the vast expanses of glyphosate spraying across the Midwest. Nobody expected resistance would show up in a small, eastern coastal state. Nobody expected the first American species to evolve glyphosate resistance would be marestail.[27] And nobody expected the prize for finding it would go to Mark VanGessel, a young assistant professor at the University of Delaware. The reaction of most weeds guys was simple: Delaware? Marestail? Van-who?

The state of Delaware is a flat, marshy stretch of land between the Delaware and the Chesapeake bays. No-tillage had a long history there, as VanGessel explained to me at a professional meeting last year. Farmers used no-till to protect the sandy soils from washing into the Atlantic. They had quickly adopted glyphosate-tolerant soybeans to reduce herbicides that were polluting the waters. Control of marestail was always erratic, but nobody cared because the few surviving plants didn't seem important. VanGessel's first phone call about poor marestail control came in 1999, three years after the GMO technology was released. Springtime conditions had been extremely dry, and everybody knew glyphosate didn't work well in dry fields. The following year, springtime conditions were ideal, but control was worse. That's when he started to look deeper and found marestail across five or six fields. Soon just about every field was overwhelmed by it.[28]

The discovery and careful follow-up research to characterize glyphosate resistance in marestail were VanGessel's ticket to whatever passes for stardom among weeds guys. Why was marestail the first to evolve resistance? He conceded as much surprise as anybody. Marestail had evolved herbicide resistance to paraquat in 1980. By the time it did the same trick with glyphosate, it had already evolved resistance to five other modes of herbicide activity.[29] We should have guessed.

Sometimes, according to VanGessel, a weed doesn't have to be especially competitive to reduce crop yields. There just have to be a hell of a lot of them. Glyphosate killed the other weeds; this gave marestail space and resources to explode. Its herbicide resistance genes spread quickly via seeds rising in wind like a plume. Farmers soon found their harvesting equipment choked with sticky, bristly stems. Within four years, glyphosate-resistant marestail was nationwide, not only in soybean fields but also in GMO corn, cotton, and canola.[30] By 2005, marestail was among the top three weeds of concern to Midwestern farmers.[31] The fluffy-seeded marestail became a signature that distinguished GMO from non-GMO soybeans.

Marestail's path to weedy success could never have been achieved without the inventions of no-till, glyphosate, and GMO crops. Two years later, glyphosate-resistant marestail populations were found in farm fields around the world, beginning in Canada, Brazil, China, Spain, the Czech Republic, and soon after in Greece, Poland and Italy.[32] And still, marestail held some surprises.

Stress Test

When news emerged that marestail had evolved resistance to glyphosate, most weed biologists assumed that a simple mutation was the cause. Just a small change in one letter of the genetic alphabet in a long sequence of DNA could mean the difference between a susceptible plant and a resistant one. That was the standard Mendelian model we had come to accept.

But resistant marestail populations didn't behave as expected. When glyphosate was applied to marestail, the active ingredient never made it to the site of EPSPS activity. Marestail captured the glyphosate and moved it into a vacuole, the cell's version of a lock-box, where glyphosate remained captive, unable to inhibit the enzyme or amino acid synthesis. This feat had never been seen before; it involved many genes and changes in plant metabolism far from the EPSPS enzyme system.[33]

There were more surprises. Some marestail plants, which were isolated far from other marestail populations, evolved resistance to glyphosate using different genetic mechanisms. In other words, they independently evolved resistance in different ways, a rare case of "evolutionary repetition." The weed that nobody had cared about was suddenly a subject of intense genetic study. Two, sometimes more, biochemical methods for resistance showed up in individual plants, or in populations of plants.[34]

The research to figure these things out helped open a previously dismissed line of inquiry: Do herbicides, themselves, cause mutations that make plants resistant? The answer is no, when normal lethal rates of herbicides are applied. After all, dead plants, mutated or not, don't reproduce. But many weeds are exposed to low, nonlethal levels of herbicides—at the edge of the field, or from spray drift. These low doses stress the plants without killing them. The herbicide might not cause a mutation, but stress can do strange things.

Stress can cause epigenetic changes in plants.[35] These changes don't alter the DNA sequence of a gene. They change the chemistry surrounding the DNA, and that alters the way a gene works. Epigenetic changes affect the ways genes are regulated—whether the gene is turned on or off. No mutations are necessary. This could include genes involved in processes that lead to herbicide resistance.

Epigenetic changes—not just mutations—can be inherited over generations and spread in a population of weeds. Evolution turns out to be more complicated than Mendel's pea experiments suggested. There's no proof that specific epigenetic changes caused herbicide resistance in marestail. But resistance has evolved following repeated low rate applications of herbicides.[36] Epigenetic changes would explain how different marestail populations ended up with different mechanisms of resistance.[37]

What Freedom Looks Like

For conventional no-till farmers, the evolution of glyphosate resistance in marestail populations was one more technology-driven punch in the stomach. The first appearance of uncontrolled marestail in a field would be understandably disconcerting. Maybe the sprayer operator had goofed up.

Maybe rain had washed the herbicide off. Maybe the soil was too dry, too wet, or too cold. A farmer's first response was to apply more herbicide, maybe at a higher rate. This was the best gift of agrestal selection farmers could have given to advance the evolution of glyphosate-resistant marestail populations: greater selection pressure helped eliminate any remaining susceptible marestail genotypes.

Once a GMO soybean or corn field was full of resistant marestail, it was too late for common-sense resistance management practices like crop rotation and occasional tillage that would have avoided resistance in the first place. Farm advisers looked for a technological solution, usually another herbicide. Surely, one was available from the dealer that sold the miracle seeds and chemicals. Farmers returned to some of the old standard high-rate, high-residue, high-leaching herbicides. The ones glyphosate-tolerant crops were supposed to remove from the landscape.[38]

Resistant marestail made its appearance at a bad time. Farmers who had adopted GMO crops with "Freedom to Farm" were soon captive to low commodity prices and oversupplies. The government had promised that free trade would open new markets and stabilize prices. But farmers in Brazil also added GMO beans to global supplies; India was not far behind. Government farm programs benefitted large producers, while low grain prices benefitted grain buyers. By the time resistance evolved in marestail, Monsanto and other chemical companies were buying up seed companies to gain control of crop genetics. Agricultural industry became increasingly consolidated and vertically integrated. The whole value chain was in the hands of production input suppliers and grain purchasers. They were one and the same. Independent, freedom-loving farmers served the conglomerates, purchasing only their inputs and selling at only their prices.[39]

The first few years of GMO crops gave farmers mixed results. Most bought the whole package of no-till, glyphosate, and GMO seeds at premium prices. Farmers used less pesticide, initially. But to get the most out of their investments they needed high rates of fertilizers, insecticides, some nonglyphosate herbicides, and spray additives. All from the same dealer. Falling land values and unstable commodity prices left farmers cautious. The only way to increase income was to get bigger and plant more. Mid-size farm operations, under the shadow of the crisis of the 1980s, found it hard to stay in business.[40] When marestail showed off its resistance to glyphosate, the cost of weed control again increased. White puffs waved in the wind above abandoned land.

Bottom Line

On a midsummer afternoon, Chet called from the Seed-N-Feed. I stopped by to confirm identification of the latest suspected resistant marestail population. The Seed-N-Feed was busy, dusty, and thick with the odor of farm chemicals. He tossed a marestail plant on the counter. Then he pulled two limp specimens from a bucket in the corner—common lambsquarters and giant ragweed. These, too, had evolved resistance to glyphosate. But the marestail plant wilting on the counter was from a field where glyphosate had not been sprayed. It had apparently evolved resistance to another type of herbicide.

Chet was in a fix: when his guys sprayed a herbicide on a field and marestail didn't die, the farmer expected him to compensate the expense and lost production. He leaned on the counter and sighed. The herbicides and seed genetics were the most advanced ever. And marestail was harder to kill than ever. His customers couldn't diversify their rotations to delay resistance because they didn't make money on crops like wheat, barley, or oats. They couldn't plow and cultivate out the weeds because they'd sold that equipment years ago. Neighbors weren't neighbors; they were folks you didn't recognize, folks with an eye on foreclosure prices of your tractors and combine. Crop farming wasn't about stewardship of the earth anymore; it was about technology, the bottom line, keeping shareholders happy.

Chet turned toward the gravel lot to watch a heavy truck lumber onto the tarmac. He tapped a pencil on an invoice pad, and I sensed there was more on his mind. "Look," he said finally, "we don't really want to spray all these pesticides on GMO crops. We've gotta do it to feed the world."

I drove away from the flag-topped silos. Farming had always been considered hard work. And now fewer people would have to do it. Fewer would have the privilege. Fewer farm families to watch over and care for the landscape. Maybe advances in technology were always this way. Long ago, most people in America were farmers; today about 1 percent are. The rest are teachers, mechanics, and engineers. Maybe. Other folks leaving farming now were Walmart associates, grocery baggers. It was no secret, some were opioid and meth users.[41] Nobody could say how far this could go, whether the freedom proclaimed by the flag atop Chet's silo could survive when all the farmland was in the hands of a few people. And feeding the world? That was a myth left over from the days after World War II. Today, American farmers feed the conglomerates.

I stopped at the light in the center of town. The window at the Hardware announced a sale on plastic buckets and overalls. Next door, the IGA grocery store had gone out of business. I drove down to the gas station. As the pump whined on, I leaned on the car, looking off toward the uniform crops that stretched out from Main Street.

The Measure of a Weed

In early 2003, I prepared for my session at the Innovative Farmers meeting. I showed results from research on managing weeds without herbicides in no-till soybeans using cover crops. A heavy stand of the right variety of cereal rye, if managed correctly, would suppress weeds. It was an attempt to link conventional no-till practices with organic practices like cover crops.

As soon as my talk was over, good old Vince, from the alternative farmer group, stood up. He wanted to talk about GMO crops. I suppressed a groan. He said seeds of "GMO weeds" like marestail were blowing into his fields. They put genetically modified genes into his field. Now his soil would no longer be organic. Vince called it "genetic debris. It's like acid in the rain, herbicides in the water, and now engineered genes and mutated weeds. The way we farm, we don't need that stuff in our lives." I wondered if I was supposed to reply.

Eli was there at the back of the room. He and Vince always seemed to work in tandem. "All those numbers and statistics you showed are fine," he began, "but I wonder if you've ever looked at the effect of the moon." I didn't want to challenge him, so I explained why it's difficult to conduct that kind of research with all the variables involved. His answer was quick: "It should be easy; there's a full moon every month." Before I could respond, he continued: "You measured numbers of weeds. That's what you measured; that's what you found. Maybe you're measuring the wrong thing. If you till weeds following a lunar cycle, you'll find other things if you measure them."

I didn't know where to begin to respond. Luckily, the session was over and nobody else had a chance to make me look more foolish. At the next session break I slipped out and drove home, trying to sort it out. These were smart guys. They held to old ideas and practices, but they were resilient farmers, or they wouldn't still be in business. They disliked conventional technology, but they were more like conventional farmers than they realized. They looked at the world differently and took different approaches toward the same end. Both approached farming with a belief in the fundamental value of their trade.

Maybe Eli had a point. What we measured was what we found. Maybe there was a mechanism connecting lunar cycles to weeds and crops. We wouldn't know if we didn't measure it. And if all we measured was weed control, we'd miss the herbicides going into the groundwater. If all we measured was weed seedling emergence, we'd miss the impacts herbicides have at the population genetic level, like evolution of herbicide resistance. If all we measured was crop yield, we'd miss everything going on in the farm community. Would it make sense to study weeds and measure happiness, community resilience, or life satisfaction?

On the Shelf

Marestail, symbol of misplaced hopes of no-till farmers, crossed the landscape to celebrate its remarkable ecological success without regard to philosophical perspectives of the soil. Alternative farmers took it as just one more weed. With their fall tillage and rotations, it was absorbed into their system. There are no cases of marestail, or any other weed, evolving resistance to tillage, rotations, cover crops, or other practices on alternative or organic farms. In fields managed by their conventional farmer neighbors, marestail evolved resistance to six herbicide modes of action, including cases of multiple resistance.

From the view of a weed that nobody had cared about, marestail left a legacy that endures from grocery store shelves to rural towns across the Midwest. The elegant biochemistry that made GMO crops work, and cleared the path for marestail's rise to fame, also propelled the anti-GMO movement, the largest global consumer campaign in history, in opposition to the technology. Driven by emotion and marketing, and less by science, the anti-GMO movement, with a huge assist from marestail, helped put non-GMO labels on everything from apples to baking soda to shampoo. The movement gained momentum when marestail (and eventually other species) evolved resistance to glyphosate after experts said that was impossible. Industry had predicted that farmers would use less herbicide by switching to GMO crops and glyphosate. But when marestail evolved resistance to glyphosate, farmers returned to the old herbicides and high application rates. Meanwhile, the move to GMO crops that were supposed to make marestail control easier accelerated farm foreclosures and industry consolidation. Thousands of farm communities, already in decline, were reduced to empty shells with antique shops and a gas station, as local and family merchants went out of business.

I walked Main Street in Smithville recently. The old Hardware was dark inside, windows were empty, a yellow notice covered the door. The IGA grocery had turned into a second-hand clothing shop. Down at the Seed-N-Feed, Chet was steadfast. The high-fructose syrup and ethanol markets were moving his corn. Cheap grain made great feed, so the animal feedlot guys were happy. His new silo was up and running, the flag flying higher than before. He had set aside an old grain bin to accept non-GMO crops. I asked how things were going with marestail. He was almost nostalgic: "Look around; it's still out there. But more resistant ones are coming, getting started across the county line. I'll give you a call when I find some specimens."

Palmer amaranth, *Amaranthus palmeri*.

6

PIGWEED

Air hanging over the primeval floodplain of the Mississippi River, in the area that locals would come to call the Delta, held omens that discouraged human encroachment. Early native and later European wanderers avoided the dense wet forests of oak, gum, cottonwood, hickory, and cypress. Thickets of giant swamp grasses lined shallow streams seeping toward the river. The muck abounded with animals and timber. It teemed with decay, disease, and hot swamp gas.[1]

Somewhat contrary to its name, the Delta is not the southern reach of the river near the Gulf, but a flat meandering plain bordering what is now Arkansas and Mississippi and reaching into southern Missouri and western Tennessee. Over thousands of years the water course has swayed with on-and-off flood cycles that moved, sifted, and deposited layers of fine-grained sediments. The rich sandy loams are fertile and bountiful. Patches of sticky-clay gumbo soils can suck the heels off your boots. On dry ground, small round aggregates at the surface give the soils their name: "buckshot."

Today, the Delta sustains one of the most intensively farmed and productive agricultural regions of the world. Large-scale, mechanized, low-labor, high-capital enterprises of cotton, soybean, rice, corn, and vegetables stretch across over 1.6 million hectares (4 million acres) of near level land. Over the last decade, the Delta has become the geographical heart of overlapping distributions of the nation's most troublesome and genetically confounding weeds, a mix of ancient, modern, wild, and mutant plants of the Amaranth family: the pigweeds.

I first visited the Delta in the mid-1980s, on an early career journey to get acquainted with the weeds, people, and agriculture of the South. I crossed the Mississippi River on Route 40 from Tennessee into Arkansas over the "Big M," the double-arched Hernando de Soto Bridge. Jumping out at the Welcome Center, the subtropical humidity filled my lungs as I adjusted to the stifling heat. It was early morning, and fog held fast in the motionless air, a cloud stretching toward an ambiguous end.

There was an undeniable splendor to the land, its horizonal greenness, regularity, and endless rows of field crops. A rising sun hit the dark sandy-loam as I stepped into fields to take in the essence of this particular place. I wondered what it had looked like to those early wanderers and what they would sense in the breeze above this landscape today. Surely, they couldn't have imagined the agricultural behemoth it has become. Along with its beauty rests an uneasiness, a sense of things unsettled in a terrain underlain by ancient canals, remnants of lost civilizations, vestiges of enslavement, and untold stories of hardship and brutalities.[2]

It didn't take long to find the pigweeds. They thrived along roadsides and field edges. There was a mix of prolific, promiscuous, and hybridizing species, difficult to differentiate one as genetically distinct from another. Some had erect, coarse stems and bristly seed heads; others flaunted tall, gangly spikes that moved drowsily in a light wind. A few seemed more modest, hugging the earth in dense patches around waterways, drainage areas, and irrigation systems.

The Delta resembles no other farming system, a landscape transformed into factory. The scale, efficiency, and skillfulness of operations figure in the precision, uniformity, and orderliness of fields stretching across the horizon. I couldn't help noticing the vigor and cleanliness of the crops. The herbicides used there were remarkably effective, all the weeds were under control. Except for a few. And those had no competitors—except the crop—so they, alone, had the run of the fields.

Beginning around 2012, the Delta became the focus of a biotechnological eruption. Uncontrollable pigweeds began to spread from the Delta to other soybean, cotton, and corn producing regions of North America. And in a botanical blink, they were "the most widespread, troublesome, and economically damaging" weeds in crop fields on the continent and into South America, the Middle East, and Australia.[3] No plant species can travel the path from wild to waif to worldwide menace without significant assistance from coevolutionary partners plying advanced technology against them. And pigweeds turned out to be especially suited to respond to mixing, stacking, and repackaging of genes and chemical weed killers that would shake the agricultural and social fabric of the Delta and beyond.

That Can't Be My Story

"Pigweed" is an all-purpose name for a handful of species in the *Amaranthus* genus of the Amaranthus family (the Amaranthaceae).[4] These warm-season plants with simple spoon-shaped leaves feature a coarse bristly seed head. This flower- and seed-bearing spike dries brown and persists long after the rest of the plant has senesced. Hence the name *Amaranthus*, meaning "never dies." *Everlasting*. The common name is less enchanting—nitrogen-loving weeds that thrive in hog lots and pig manure.

Pigweeds originated in canyons, riverbanks, and tidal marshes. Few species tolerated such drought, flooding, and intense heat. Natural selection favored genotypes that produced many small, dark, lens-shaped seeds easily dispersed by water, moving soil, passage through the gut of birds and other animals. These plants would thrive where humans scratched the earth to grow crops. Ancient people knew them for their edible seeds and fleshy leaves.[5]

I will distinguish two types of pigweed: "classic" pigweeds, familiar to farmers and gardeners, have had a long and close association with humans. "Modern" pigweeds gained attention only recently. Even during the first few centuries of European-style agriculture in North America, the modern pigweeds stayed in canyons and creek beds until farmers coaxed them into fields of industrial crops and then pounded them with advanced chemical farming practices.

Classic pigweeds came from North America and moved to the Old World in the mid-1700s. Most recognizable are redroot (*A. retroflexus*), smooth (*A. hybridus*), tumble (*A. albus*), and prostrate (*A. blitoides*) pigweeds. Peter Kalm, the Swedish botanist, collected pigweed seeds in Pennsylvania (with possible assistance from Ben Franklin and John Bartram) and sent them to Carl Linnaeus. Good old Linnaeus grew them, classified them as *Amaranthus retroflexus*, and sent seeds to his buddies at botanical gardens around Europe. Thus, pigweeds went through the hands of celebrated naturalists on the way to inoculating an entire continent with one of its most pernicious weeds. By 1783 pigweeds were in Paris and by 1800 farmers across Europe were hoeing and pulling them from farms and gardens. Within one hundred years they had spread to Norway, around the Mediterranean, and into northern Africa and the Middle East.[6]

Classic pigweeds are mostly self-pollinated. There are separate male and female flowers arranged on the same plant. Botanists call them "monoecious" ("one-house"). Pollen usually falls from a male flower of the plant to a female flower to make seeds. Wind and insects move pollen around, allowing different classic pigweed species to hybridize (cross-pollinate).

Until about twenty years ago, redroot, smooth, and other classic pigweeds represented the principal Amaranthus contribution to weediness in much of the world. It was a significant and widespread contribution. They achieved a modicum of ecological success and were recognized as annoying to farmers and gardeners, who have been hoeing and pulling them for centuries.

Modern pigweeds also came from North America. Palmer amaranth (*A. palmeri*) evolved in canyon stream beds from Texas to Southern California and into Mexico. This is where "Palmer" remained for centuries.[7] Waterhemp (*A. tuberculatus*), another modern pigweed, evolved along river edges from Nebraska to Indiana and Ohio. Farmers had noticed waterhemp before 1800, but only as a mere curiosity.[8]

Modern Palmer and waterhemp, unlike their classic cousins, must cross-pollinate. They have only male or female flowers on separate plants.

Botanists call them "dioecious" ("two houses"). Self-pollination is impossible. This obligate outcrossing assures a high level of genetic mixing. Pollen becomes airborne and can travel a few hundred kilometers before settling down to share genes with random, unknown female strangers.[9] The two moderns, Palmer amaranth and waterhemp, can hybridize, sharing pollen with each other to make plants with genetic combinations unlike the mother or father plant. Strangely, Palmer and waterhemp can also share pollen and hybridize with the classic pigweeds. This cross-subspecies feat is quite unusual and makes it difficult to determine with confidence the identity of pigweed specimens. The tall floppy spikes of Palmer, waterhemp, and their variable hybrids produce thousands of small, green, inconspicuous flowers and millions of seeds.

A decade into the twenty-first century, modern pigweeds began to slither out of creek beds and canyons. Agricultural expansion had put more land into cultivation and moved the different pigweed species within proximity for genetic mixing that would not otherwise have occurred. With cross breeding, pigweeds gained novel sets of genotypes on which agrestal selection (weed evolution in response to agricultural activity) could work to favor plants best adapted to agricultural field conditions. Pigweeds became a highly variable cosmopolitan weed syndrome that persists as a hybrid swarm of classic and modern pigweed genes.[10] The modern outcrossing breeding system might have made mixing of pigweed genes inevitable over time. But that's only a start. The rest of the pigweed story belongs to the genetic tricks of modern Palmer amaranth, waterhemp, and their human accomplices.

Keeping Careful Track

Pigweeds were targets of early chemical weed killers. Chemists of the 1800s teamed up with farmers to concoct solutions of copper salts, iron sulfate, and copper nitrate to kill pigweeds in grain fields. More complicated thiocyanates, dinitrophenols, arsenicals, and others came into use in the early 1900s.[11] These toxic brews were difficult to handle, had to be applied at high rates, and could injure crops—and people—as much as weeds.

The weed killer 2,4-D came along in the late 1940s and promised to make farming easier for American soldiers returning to farms after the war. Farm magazine stories and farm visits by agricultural extension agents taught farmers to spray 2,4-D over corn fields to kill pigweeds and other broadleaf weeds without killing the corn. The commercial success of 2,4-D launched the herbicide industry. Every year, new products were put on the market to kill different weeds in different crops.

A widely useful family of herbicides called triazines became available to farmers in the late 1950s. Products containing triazines were given technical-sounding names (e.g., atrazine, simazine) befitting the era of scientific chemical farming. Farmers learned about them by word-of-mouth from neighbors or during visits to the local extension office. There was little need for marketing; they were the only products around that killed so many different weeds.

Pigweeds didn't stand a chance. Farmers sprayed triazines to kill pigweeds and other broadleaf weeds along with some small-seeded grasses. These herbicides seemed to fulfill the dream of humanity since the dawn of the Agricultural Revolution. Crops, not weeds, would cover the fields. With the help of herbicides, farmers would put away their dusty mechanical weed cultivators. They would stop rotating to less profitable crops to break up weed life cycles. They would stop turning the soil every year to bury weed seeds. They would put an end to long hours in the sun moving slowly across fields to scrape weeds out of the soil. All farmers had to do was spray. And spray some more.

Little wonder that chemical weed killers became the dominant approach to weed control by the 1970s. Competition in the industry was tough. So were the product names, like "Marksman," "Bullet," and "Prowl." Advertisements featured clenched fists and predatory animals. The chemicals stayed active in the soil for weeks or months to kill weeds that started to germinate into the growing season. Six weeks after corn planting, hardly a weed in the field, and no labor wasted on mechanical cultivation. Few things could be so satisfying.

After twenty years of suppression by triazines, the everlasting pigweeds reappeared. In fields where these herbicides had been used in continuous cropping, farmers started to see occasional pigweed plants that had not been killed by the standard treatment. At first, farmers weren't sure what was happening. There were many reasons weeds might escape a herbicide treatment. Sprayer glitches or unfavorable environmental conditions allowed small patches of weeds to pop up here or there. A few floppy-leaved pigweeds didn't draw much attention.

But these escaped weeds were different. Repeated spraying of triazines throughout the Corn Belt had imposed severe selection pressure on the pigweed populations. Normally, pigweeds were killed by triazines. Now, throughout the 1970s, more and more mutant pigweeds survived and started to take over the fields. They grew more abundantly, produced more pollen, and spread their genes more widely than the other pigweeds. In just a few growing seasons, none of the ordinary herbicide susceptible pigweed plants remained. Pigweeds became the first major weeds to evolve herbicide resistance: heritable genetic changes that allow weeds to survive and reproduce after receiving a dose of herbicide that should otherwise have killed them.[12]

Farmers didn't notice this until harvest time the following year. They put the crop in the ground and applied herbicides. The corn should grow, the weeds should die. By the time the resistant plants were noticed, the corn was too tall to get equipment in the field to do anything about the weeds. The pigweeds sucked up the water, fertilizer, and other soil nutrients, and then shaded the crop so it couldn't get enough light to produce normally. The bristly spike of herbicide-resistant pigweeds produced many thousands of seeds. The next spring pigweed covered the ground, most of them triazine-resistant.

Everybody knew the problem would soon be solved. The solution would be waiting on the shelves at the local herbicide dealer. A new family of herbicides called ALS-inhibitors came on the market in the 1980s. Repeated spraying of triazines had imposed selection pressure that led to evolution of triazine-resistant weeds. The obvious response was to reduce use of traizines and switch to repeated spraying of ALS herbicides. The industry reps and university extension folks were ready to help.

ALS herbicides were marketed aggressively. Industry-sponsored meetings featured ag-extension presentations with seemingly endless figures and bullet points using a new technology called "PowerPoint." Product-label signs sprouted along rural highways, promos filled time on farm radio, and semi-technical articles appeared in farm magazines.

The message of ALS promotion emphasized environmental gentleness along with product effectiveness. This would counter the bad press following herbicide contamination of waterways and drinking water. Macho names were out. ALS herbicides were kind to nature. They had names like "Beacon," "Harmony," and "Accord." Advertisements featured young women in light dresses pumping (presumably pesticide-free) water at the well while cows grazed serenely on the hillside.[13]

Pigweeds didn't stand a chance. ALS herbicides were remarkably effective. With doses as low as 2 grams per hectare (3/4 ounce per acre)—less than a thousandth of the amount needed for the old products—pigweeds stopped growing, turned purple, yellow, crispy, and died. Farmers adopted ALS herbicides quickly to clear weeds out of corn, soybean, cotton, and wheat fields. There were peaceful ALS products—sometimes several—for every major crop, and they could be applied before and after planting. Farmers would use one or more of them in every crop, every year. Those triazine-resistant pigweeds popping up all over the place would soon be annihilated in one crushing blow of botanical brutality.

All Will Be Well

After five years of suppression by ALS herbicides, the everlasting pigweeds reappeared. At first, farmers weren't sure what was happening. Most of the pigweeds stopped growing, but by the time they were dead, other pigweeds were already taller than the crop.

Something about these escaped plants was different. These were tall, robust, pigweeds with long taunting spikes. In just three years, modern Palmer amaranth and waterhemp populations evolved resistance to two types of ALS herbicides. It began in Kansas in 1993. Two years later, resistance to five types of ALS herbicides was discovered in Arkansas populations of Palmer amaranth. Across the border in Missouri was waterhemp with resistance to ten brands of ALS herbicides. Even sprayed a thousand times the normal amount, some of the plants still survived.[14]

In a heartbeat of botanical time, all pigweed species became suspect. Palmer and waterhemp, hardly known outside their native habitats, were suddenly the focus of weed control efforts across major corn, soybean, and cotton production areas. Farmers were particularly alarmed to find ALS-resistant waterhemp growing in fields where ALS herbicides had never even been applied. This seemed impossible. Without selection pressure from repeated ALS herbicide spraying, ALS resistance could not have evolved. How did ALS-resistant biotypes get there?

Pollen, of course, moves in the wind—like a virus, only farther. ALS resistance had evolved in distant fields. The mutation for the altered ALS enzyme was carried in pollen produced by resistant male plants. Female waterhemp plants within reach of airborne pollen were fertilized by that pollen with the genes for ALS resistance. These pigweed mothers produced seeds—thousands of them—with the mutation for resistance to ALS herbicides. Suddenly every field across the continent, and ultimately across the world, had the potential to explode with resistant pigweeds wherever ALS herbicides were used.

In the mid-1990s I started to get reports of weeds in Ohio that had escaped control by ALS herbicides. Growing conditions and selection pressure from ALS herbicides favored evolution of resistance anywhere across the state. Waterhemp was not common in Ohio, but isolated populations dotted the Ohio River Valley. The landscape was fragmented into patches of fields and lakes and forests and urban zones. But the airscape linked whole ecoregions, their communities, and farm fields. So Ohio farmers were linked to Arkansas, Delta, and Kansas farmers by pollen floating in the air. I figured it was only a matter of time before pollen containing genes for ALS resistance would move along with robust pigweeds to the eastern edge of the Corn Belt.

Everybody knew the problem would soon be solved. It would be waiting on the shelves at the local herbicide dealer. The new, genetically modified (GMO), crop seeds with tolerance to glyphosate herbicide (i.e. "Roundup Ready") were going to put an end to weed problems. Repeated spraying of ALS herbicides had imposed selection pressure that led to evolution of ALS-resistant weeds. The obvious response was to reduce use of ALS and switch to repeated spraying of glyphosate.

A slick new marketing campaign promised a new era of GMO crops that would be more productive, easier to grow, and gentler on the environment. Rural radio and local TV advertisements and "infomercials" featured company technical reps. The same guys appeared at industry-sponsored promotional presentations with snacks and hats with colorful logos. New signs grew along roadsides beside fields of "Roundup Ready" soybeans, not a weed in sight. Famers across America would be first to audition genetic modification technology over millions of hectares of cropland. Who wouldn't want to be a part of that?

Farmers rushed to plant the GMO seeds throughout North and South America and Australia. Then rushed to spray glyphosate across millions of hectares of soybean, corn, cotton, and canola crops. Some sprayed it several times per year. Glyphosate was familiar and easy to use; farmers had been applying glyphosate in other ways for two decades.

In just a few years, GMO crops dominated the landscape. You could pick those crop fields out from miles away: they were the ones with no weeds. The scale and extent of those clean fields grew along with the success of glyphosate around the world.

In Light of This Logic

For weeds, as for much of life, timing is everything. Germinate at the wrong time—too close to winter, too deep in the earth—and you'll never make it. Flower at the wrong time—too early or too late—and few or no seeds will result. Either way, the opportunity to pass your genes on through generations is lost. The ability of weedy plants to pop out of the soil in springtime—soon after the previous batch of seedlings was hoed out—is a frustration to gardeners everywhere. This timeliness is the outcome of thousands of years of unconscious agrestal selection that advanced pigweed adaptation to regular cycles of agricultural activities. Favored genotypes were those adapted to these cycles and to all types of habitat disturbance, including seed burial with plowing, movement with cultivation, and dispersal with harvest.

Pigweed seeds fall off the plant in late summer or early fall. They are initially dormant, so they sleep out the winter. When soil warms again in springtime, they still won't germinate unless they're close enough to the soil surface to make a go of it. Some wait many years before they find themselves in a place and time favorable for growth. How do they know when the time is right to germinate?

Pigweeds evolved clever mechanisms to sense their environment so germination can proceed when conditions are favorable. They sense and respond to light—wavelengths at the edge of the visible range. Pigweeds' light tricks are controlled by a set of proteins called phytochromes. The phytochrome enzyme acts like a switch to activate key processes like germination (also flowering). Red light turns the switch on and promotes germination; far-red light turns the switch off and inhibits germination. When seeds remain buried deep in the soil for a long time, the phytochrome switch gets turned off. This way, pigweed seeds don't accidentally start to germinate when they're too deep in the ground. The seeds wait to receive a signal—a brief exposure to red light—telling seeds they are close to the surface, a safe time and place to germinate.

When farmers and gardeners till the soil to prepare land for planting or to control weeds, pigweed seeds are brought to the soil surface. If seeds receive a flash of light before they fall on or into the soil, the phytochrome switch gets turned on, germination can proceed, and the seedlings begin to grow. Tilling the soil again, to kill that batch of weed seedlings, exposes another bunch of seeds to light and sets off the next flush of germination. With thousands of pigweed seeds in the soil, this process can repeat itself throughout the growing season.

That is what inspired my research group, in the 1990s, to fumble around in the dark on moonless nights. If exposure to light during tillage caused pigweed seeds to germinate, we reasoned, maybe tilling the soil at night—without light exposure—would keep pigweed seeds from germinating. Any seeds exposed in darkness and reburied by tillage would not see any light; they would think they were still buried, stay dormant, and refuse to germinate. We would keep the phytochrome switch turned off.

The only way to test this theory was to convince a couple of graduate students to conduct field and laboratory experiments in the dark. In autumn, we buried small mesh bags of freshly harvested pigweed seeds. The following May and June (pigweed germination time) we recovered the buried samples on dark nights. Meanwhile, another student fired up the tractor to drag a disk cultivator across the field and till the soil at randomly assigned places, no headlights or taillights shining. Plots tilled at night were paired with plots tilled in full sunlight the following day.

The seeds collected at night had to be handled carefully. Even a brief exposure to dim light could activate the switch. The seed bags pulled from soil at night were immediately put into dark jars, sealed with black tape, and shuttled to the dark windowless lab. Still in the dark, we opened the bags, rinsed and separated the seeds, and placed some of them in dark boxes for germination treatments—a range of temperature and moisture conditions. Other seeds were spread out on moist paper and exposed to specific wavelengths of light for exactly one second. If you accidentally opened the light box—even a crack—the flash of light burned on your retina and left you even more disoriented.

My musty lab in the basement was the perfect spot. It was a creepy place to spend hours in the dark of night. There were banging pipes, hissing valves, and scratching in the walls. Working in darkness, we manipulated petri dishes, filter paper, tiny seeds, and tried not to knock over the glassware or flasks of nutrient solution. We worked quickly to get out of there. And returned the next day hoping to find the germination chambers closed tightly so no light could enter.

The results were fairly clear. Fewer pigweeds germinated and emerged in the field when tillage was done at night rather than during daylight. Lab experiments confirmed that phytochrome was most receptive to low light flashes in springtime when soil was warming, and seeds had imbibed sufficient water. The differences were not great, but we had expected results to be complicated by environmental conditions and different photoreceptors that are activated at different times during germination. Still, we fooled about half of the seeds into not germinating when soil was tilled at night, even when conditions of temperature and moisture were right.[15]

Secret night missions to undermine the pigweed conspiracy were a small part of my research to find ways for farmers to manage weeds without continually reaching for the next jug of toxins. In other experiments we tested cover crops to suppress weed growth, plant residues that released weed seed germination inhibitors, beneficial insects and fungi that we hoped would destroy weed seeds. Many years before drones, we flew model airplanes over crop fields to capture images of weed patches. Along with colleagues across the north central United States, we tinkered with offbeat ways for farmers to reduce the use of herbicides. There were experiments on things like microwaves, sandblasting, heatwaves, steam, soil solarization, boiling liquids, plant extracts, insect venom, vibrating rods, manure teas, seed leachates, stale seedbeds, anaerobic disinfestation, fungal toxins, weed seed capturing devices, and robots of various description.

Pigweeds were a model warm-season test species for these studies. You could always count on them being there—the soil seedbank was full of them. They were recognizable and reliable, even if they weren't (yet) on anybody's list of the most important species. And some of the things we tried actually worked. Like night tillage. Well, sometimes—under the right conditions.

We knew that conventional large-scale farmers might not be inspired by these approaches. With tight profit margins, uncertain commodity prices, and increasing farm sizes, farmers needed consistent weed control to get a good yield and stay in business. They weren't about to take a risk on unfamiliar and unreliable practices that were hard to implement on a large scale. By tweaking weed management tools, we wouldn't transform the agricultural system into something it was not. Manipulating nonchemical technologies to shift weed populations around was a subversive way to challenge the forces driving weed resistance.

From the perspective of most farmers, glyphosate and other herbicides were working just fine. So fine that by the early 2000s, more than 136 million kilograms (300 million pounds) of herbicides, about 0.45 kilogram (1 pound) for every person in the United States, were applied on GMO corn, soybean, and cotton fields. A custom applicator could spray the newest herbicide products across hundreds of hectares. Spraying was quick, reliable, and—if done correctly—effective on a big scale. The logic of chemical farming was convincing: products that attack a specific plant enzyme can reliably kill 99.9 percent of the target weeds.

No allelopathic residues, spit coffee grounds, or robotic tillers could match the continual flow of new herbicides. Besides, there isn't much profit in things like smother crop seeds. Nobody—except maybe a sleep-deprived farmer—makes money when tillage is done at night. The logic of testing alternative approaches rested in rural tensions and in uncertainty about what was happening with the other 0.1 percent of those weeds. The mutant ones that were not reliably killed.

Rumbling Like Thunder

After about nine years of suppression by glyphosate in GMO crops, the everlasting pigweeds reappeared. Glyphosate-resistant Palmer evolved in Georgia cotton fields in 2004, followed quickly in Arkansas, then throughout the South, and spreading northward to Ohio and Michigan. There was glyphosate-resistant waterhemp, too. First in Missouri, and soon in twenty US states from Louisiana to North Dakota and north to Ontario. Some populations were resistant to five or six classes of herbicides.

At first, farmers weren't sure what was happening. This wasn't supposed to happen with glyphosate.[16] You drove by in the truck one day and didn't see them. You drove by the next week and they were 30 to 40 centimeters (12 to 16 inches) tall. Once pigweeds reached a height of about 10 centimeters (4 inches), growth took off, and you couldn't stop them. Cut them down to the ground at that stage, and new stems would simply shoot back up from buds at the base of the plant. A single escaped plant could produce 60,000 to 100,000 seeds, enough to infest over a hectare (about 3 acres) in two years, and 18 hectares (45 acres) three years later.

The resistant Palmer populations were unusually dense and competitive. Farmers who once produced 3,400 to 4,000 kilograms of non-GMO soybeans per hectare (50–60 bushels per acre) now harvested only 670 to 1,000 kilograms (10–15 bushels) where the resistant pigweeds grew. By 2011, waterhemp had moved into more than 1.2 million hectares (3 million acres) of farmland where it could cut crop yields in half.

Some farmers sent teams of laborers, called "choppers," into the fields. Choppers walked the rows of cotton to chop out pigweeds with a hoe or cutlass. It was an expensive last-ditch effort. If the pigweeds were flowering, choppers had to collect the plants and drag them out of the field to prevent resistant seeds returning to the soil. The mature pigweeds towered over crops. At harvest, the massive combines got choked and stalled where dense patches of thick pigweed stems clogged up the machinery.

Farmers had few alternatives. They knew only chemical control, precision-applied by a computerized rig with a 46-meter (150-foot) boom that could blow across 1,125 hectares (2,781 acres) in a day. Some switched back to old products and increased the dosages. Glyphosate-resistant pigweeds stifled hopes that GMO crops would reduce chemical usage and put an end to herbicides finding their way into the Mississippi River. Farmers had little experience with slow, clunky mechanical cultivators. They certainly were not going to use vibrating rods or to shoot coffee grounds at robust pigweeds. Night tillage? You had to be kidding.

In springtime, large-scale American farms took on operating loans of tens of thousands of dollars. They had relied on GMO crops and glyphosate to control weeds and guarantee good harvests. But now, glyphosate-resistant pigweeds were everywhere. Crop yields suffered, harvest was delayed, and more fuel was needed to dry the crop contaminated with pigweed residue. The extra money for GMO seeds was wasted. Farm magazines covered stories of farmers who had abandoned fields infested with weeds they could not manage. The big operating loans couldn't be repaid. If they were lucky there might be a flood or a drought, and crop insurance would get them by. Some went out of business.

Mutations of Immortality

With glyphosate resistance running through pigweed populations, nothing could stop the expansion and dominance of Palmer amaranth and waterhemp. By 2014, it seemed that pigweeds had reached a pinnacle of ecological success. They had become the most significant widespread weeds, and shared most of the landscape, centered around the Delta. "Pigweed" no longer referred to your grandmother's garden-variety redroot or smooth nuisance. To farmers, researchers, and the chemical industry, "pigweed" was now the modern, outcrossing, hybridizing, dioecious pest syndrome synonymous with industrial biotech agriculture.

We knew that pigweeds could spread herbicide resistance genes in airborne pollen. We knew from other weeds, like marestail, that glyphosate resistance might be caused by mechanisms that prohibit movement of the herbicide to the active site of enzyme activity. But modern pigweeds played a different genetic trick called gene amplification. Glyphosate kills weeds by inactivating a plant enzyme called EPSPS. Resistant pigweeds amplified—made hundreds of duplicates—of the gene that makes EPSPS. The plants made so much EPSPS enzyme that glyphosate could not inactivate all of it. It was impossible go get enough glyphosate into the plant to inhibit all the EPSPS enzymes, so the gene-amplified pigweed plants survived.

Waterhemp played yet another trick. The amplified copies of the EPSPS gene were not scattered throughout the plant's DNA as researchers expect. Instead, glyphosate-resistant waterhemp created an entirely new chromosome to carry the amplified EPSPS gene. More surprising, the extra chromosome is ring-shaped, and it is sexually transmitted during waterhemp reproduction.[17] These discoveries have been a thrill, of course, to weed biologists studying evolutionary behavior at the molecular genetic level. A lowly weed had created a whole new, nonchromosomal, heritable, circular DNA structure. To heck with slow-mo Darwinian evolution and Mendelian inheritance the way we all learned it in school. Duplicate genes on rings of DNA: and to think it was first found in pigweeds. In Kansas!

They'll Race at Top Speed

On a sunny summer day in 2016, I got a call from Chet down at the Seed-N-Feed. He was my main contact for farm supplies, especially unusual crops and varieties I needed for my field research. He wanted to send one of his crop consultants my way, hoping I could identify a weed from the adjacent county. I would be working in the field that afternoon but would be happy to take a look at another oddball specimen. Around mid-afternoon a white truck picked up dust along the road. Without doubt, the plant Chet had sent me was Palmer amaranth, the first I had seen in this part of the state. There were the characteristic long petioles and dark leaf spot. Spiny bracts on the flower spike meant the specimen was a female plant, too young, mercifully, to shed seeds into my research plots. Chet's consultant had told the farmer that the weed was probably Palmer, but he wanted confirmation. "He was hoping it was marestail. Nobody hopes they have marestail until they see this one coming."

Over the next few weeks, I scouted around for more Palmer amaranth in the area. With help from the local extension office, I found it in three GMO soybean fields. The next year, those fields were again planted to GMO soybeans. The glyphosate-resistant Palmer was taller, thicker, more ominous. I spoke with neighboring farmers. It seemed we still had a chance to keep this weed from spreading throughout the area. I suggested rigorous scouting, simple management changes, and a combination of chemical and nonchemical control measures. Neighboring farmers would work together to implement the strategy across the watershed. They looked at each other and nodded. They looked at me and smiled. "Well Doc," one of the farmers intoned, "it's going to get here eventually. We may as well face it."

I ran this by Chet. As a seed supplier and pesticide applicator for farms across the region, he was in a key position to watch out for new herbicide-resistant weeds. I described how we could work together to protect an agriculturally significant county from an especially troubling weed. Chet tapped his pencil on the counter and looked at me like I was a university professor who didn't understand the situation. He explained that growers don't believe resistance prevention measures would succeed in keeping resistant pigweeds off their farms. Plus, about 40 percent of the farmland was rented. Growers working rented land want to make money from whatever crop they think will grow best on that land. They're less concerned about how their management affects weeds in fields they might never farm again. And some landowners restrict what crops can be grown, so farmers aren't free to rotate crops even if they wanted to.

Besides, resistance was somebody else's fault. As Chet explained, most farmers believed that herbicide-resistant weeds appeared on their farms due to poor management by their neighbors. It was the neighbor who brought on the resistance problem and then let weed seeds or pollen move across the fence line. Farmers had seen this with triazine- and ALS-resistant weeds; and with glyphosate-resistant marestail. They had no reason to think they could do anything to keep glyphosate-resistant pigweeds off their farms. All they could do was to keep buying the products industry sold them.

Chet took a phone call and I looked around the shop. Colorful logos, posters, and pamphlets were covered with a light layer of fertilizer powder. The air was heavy with pesticide solvent. Chet interrupted my dusty solitude by blasting on the intercom to notify the warehouse of an order coming in and held up a finger to signal he had one more thing to tell me. "Farmers have got used to you weeds guys," he mused. "Every new weed appears, you come up with a new solution. This resistant pigweed is movin' our way. You're not going to let us down now, are you?"

Doing the Math

Everybody knew the problem would soon be solved. Crop seeds genetically modified to tolerate one herbicide—glyphosate—were good and made a lot of money; crop seeds genetically modified to tolerate two or more herbicides at the same time would be better and make even more money. Biotechnologists had figured out how to modify crop plants by "stacking" genes for different traits, one on another. Chemical companies—which were now the same as seed companies—could stack genes for glyphosate tolerance on top of genes for tolerance to other herbicides at the same time in the same seeds.

The other herbicide they chose was dicamba. Farmers had used dicamba for a long time to kill weeds in corn fields, but not in soybean or cotton fields. Now, with a gene for dicamba resistance stacked onto the gene for glyphosate resistance, both herbicides could be used. They gave the stacked-gene seeds a fancy new name—"Xtend" and "Enlist"—and ordered up some new hats with fancy new logos. Repeated spraying of glyphosate had imposed selection pressure that led to evolution of glyphosate-resistant weeds. The obvious response was to spray both glyphosate and dicamba together. No weed could evolve resistance to two herbicides at the same time. Farmers were used to paying more for GMO seeds and had been using dicamba successfully for years; what could possibly go wrong if they just used a little more—or better yet, a whole lot more?

Dicamba herbicide had been registered for use in 1967 as a more powerful version of 2,4-D. Like 2,4-D, dicamba injured all broadleaf plants, including crops (soybean, cotton). It was used mostly on grass crops like corn and wheat. Researchers found a common soil bacterium (*Strenotophomonas maltophilia*) with a "DMO" gene that could disarm dicamba. If the DMO gene was put into seeds of soybean and cotton plants, they, too, could disarm dicamba. Since no other plants contain the DMO gene, farmers could spray dicamba over "dicamba-safened" cotton and soybean fields and kill all the broadleaf weeds, while the DMO gene protected the crop plants from damage.[18] Pigweeds wouldn't stand a chance.

Advertising started years ahead of time to set the stage for massive sales and adoption. Websites and social media promoted stacked-gene technology. Seed and pesticide dealers invited farmers to promotional presentations and provided lunch and a hat and maybe a jacket with colorful logos. Everybody came away with links to websites with all the information needed to understand how to use the product. At webinars, ag-extension specialists showed results of their field trials. With Enlist and Extend seeds, farmers could spray dicamba (or 2,4-D) over the top of dicamba-safened soybean and cotton fields to put an end to glyphosate-resistant pigweeds, which by now had evolved resistance to half a dozen additional modes of action.

There was only one concern about dicamba. The active ingredient was volatile. It evaporated easily into the air. The liquid herbicide became a vapor, or gas. When herbicide vapor molecules rose in the air, they moved wherever the air moved. If the air moved dicamba molecules on to neighboring (non-safened) soybean or cotton plants, those crops could be injured. If the air moved dicamba to the neighbor's tomatoes, cucumbers, roses, or grapevines, which are especially sensitive, those plants could be injured or killed by just a whiff in the breeze. It was fairly easy to identify the characteristic dicamba injury symptoms: cupped leaves and twisted stems. Severe injury stunted the plants and reduced yields. Dead plants were twisted, yellow, and simply dead.

Luckily, farmers were already familiar with dicamba. They knew about its volatility. They knew how to avoid the risks of crop injury. They sprayed dicamba only in early spring: low temperatures kept volatility down and sensitive crops were not yet leafed out. Farmers had been using dicamba this way for about forty years. In fact, it was illegal in most states to apply dicamba after April 15. Farmers wouldn't risk the thousand-dollar fine for violations.

But pigweeds are a hot-season species. To kill pigweeds, dicamba had to be applied in May or June. By that time, temperatures were high and volatile dicamba might move in the air and cause unacceptable damage to other crops, trees, or vegetable gardens. Farmers were not willing to take that kind of risk.

The solution to volatile dicamba would soon be waiting on the shelves at the local herbicide dealer. A new, "low-volatile" formulation of dicamba had been invented to go along with the new GMO dicamba-safened seeds. Farmers had learned all about it at meetings the previous year. I saw results of some of the field research. The low-volatile dicamba formulation was applied to four rows of the GMO stacked-gene soybeans. Just 76 centimeters (30 inches) away were four rows of conventional, non-GMO, soybeans without genes to protect them from dicamba. As far as I could see, there was not a trace of injury on the un-safened soybeans. The results, in small plots, looked stunning. They would be shown across the Midwest to convince farmers that the new dicamba formulation was safe to apply on a massive scale, even in summertime.[19]

The new dicamba-safened cotton seeds went on the market in 2015 (soybeans in 2016). Farmers ordered them quickly for planting. But come springtime, the new low-volatile dicamba was not available. It was held up in court until laws were changed to allow farmers to apply dicamba after April 15 to kill pigweeds throughout the summertime.

By May 2015, the new seeds with stacked genes had been planted. Crop plants were pushing out of the ground and pigweeds were coming on. The new low-volatile dicamba could not yet be sold. But the old, more volatile, dicamba formulation was still around. And it was cheaper than the new one not yet on the shelves. Farmers had taken on large debt to purchase seeds, fertilizers, pesticides, fuel, and everything for the season. Crop prices were low, profit margins thin. With several thousand acres of beans or cotton, you could save tens of thousands of dollars by spraying the cheaper, old dicamba. The fine for illegal application was $1000. It wasn't hard to do the math.

Out of Thick Air

Herbicide volatility happens after the product is applied. Herbicide molecules resting on soil or leaf surfaces escape into the air. If atmospheric conditions are right (high temperatures, low humidity, and low wind), these molecules hang in the air together with those that escape nearby. The result, over several hundred acres of soybeans or cotton, is a virtual cloud of herbicide molecules. This isn't just chemical drift along the edge of a field, which happens when herbicides are sprayed on a windy day. A volatile herbicide fog can tumble across the landscape, over the river, into the valley, trespassing on neighbors' vegetable gardens, non-GMO soybeans, or prized French hybrid grapes.

I started getting calls around June 2015. Most were from fruit and vegetable growers with twisted plants and misshapen fruit. Would it be legal to sell or process fruit from twisted plants? Were they safe to eat? Would the entire crop worth thousands of dollars have to be destroyed? Fruit and vegetable crops were especially vulnerable; those plants are very sensitive to herbicides like dicamba. Farmers sent samples to the State Department of Agriculture, which oversees compliance with pesticide regulations. Applying herbicides in a way that is contrary to label directions is illegal. In a corn and soybean state like Ohio, pesticide complaints had been few. Regulatory agencies were not prepared for dozens of complaints from high-value fruit and vegetable growers whose crops were getting hit hard. One of the state's largest commercial growers of tomatoes told me that three herbicides had been detected in his plant samples. Someone at the state agency suggested he simply plant something besides tomatoes.

Finally, in July 2015, the new low-volatile dicamba formulation became available. It worked surprisingly well. It was much less volatile. It cleaned out the broadleaf weeds in stacked-gene dicamba- and glyphosate-tolerant soybeans and cotton. This was July—the pigweeds were already thick and tall. Acres of ugly bristly pigweeds sprayed with dicamba were left twisting and choking to their death while soybean and cotton plants pushed on unfazed. Few things could be more satisfying.

Meanwhile, in the Delta and across the South, pigweeds were reaching over the top of cotton plants. A lot of the older, volatile formulation was still around. It had been purchased early in the season. There was one easy way to get rid of it. Something about those fast-growing, beastly, robust pigweeds changed everything. Something about knowing that the solution to make farming easier again was sitting in a jug out in the barn rattled the psyche. Something about the meaning of "weed" muddled one herbicide formulation with another. Something, somehow, let loose a lot of volatile herbicide into the airscape that connects the land and its inhabitants.

Abuse of the old dicamba herbicide formulation was pervasive in (but not confined to) the Delta. Dicamba injury burned leaves, twisted stems, and aborted flowers on vast fields of non-safened soybeans and cotton, the lifeblood of Delta agriculture. It damaged peaches and ornamentals and hardwood trees. Injury was worse where temperature inversions trapped the herbicide vapor low in the air and caused the toxic cloud to spread across wide areas.

There was no telling where a herbicide cloud would land when it crept through the air across the flatness of the Delta. Crops without dicamba-safened GMO seeds started to crinkle and twist. Farmers were not strangers to crop injury from herbicide drift. It happened sometimes, even with legal applications. In the past they just talked it out, compensated damage, made right about it. After all, they were neighbors. They hunted together; their kids played soccer. They looked out for each other.

The Delta is a place both of unconscious beauty and unspoken tension. There's a long history of single-minded independence. Fields are highly managed, controlled, designed, and constructed to extract as much bounty as possible from every measure of land. Yet farmers and laborers live on the financial edge. Everything is scaled for large expanses of soil, uniform operations, redundant customized equipment, contingency planning, and repayment of massive loans. Margins are thin. A mistake across thousands of acres can be ruinous—the wrong crop variety, the wrong herbicide.

Sales of the new low-volatile dicamba soared the next year. But there was a certain problem with the new low-volatile dicamba: "low volatility" does not mean "no volatility." Even the new low-volatile dicamba could injure crop plants and vegetable gardens, sometimes miles from the site of application. The most careful herbicide applicator, working early in the morning when temperatures and volatility should be low, could not control environmental conditions during the entire time it took to spray a few hundred acres of crops. When temperatures rose and the wind stayed low, enough dicamba from the low-volatile formulation could load the atmosphere above the crop field. It might even take the glyphosate with it.

When leaves on your non-GMO soybean or cotton plants started to curl and stems started to twist, it was hard to know whose herbicide application was at fault. All you could do was call around. Most neighbors would open their books and show their application records. But now some refused. Everybody knew who followed the rules and who didn't. Or thought they knew. You sat in your pew at First Baptist during a sermon about Cain and Abel and the limits of brotherly love and wondered if maybe the fella across the aisle was the one. Accusations were made. Discussions turned to arguments. Doubt, mistrust, anger, and resentments were laid bare. There were stories of public yelling matches, scuffles.[20]

In early October 2016, a farmer from Tiponville, Tennessee, shot and killed his forty-nine-year-old neighbor in an argument involving dicamba spraying. News reports of the killing were sparse and sketchy.[21] Later that month, on a rural road near the Arkansas-Missouri border, Mike Wallace, a fifty-five-year-old farmer from the Arkansas Delta, was killed by a nearby farmworker. Wallace had been quoted in the *Wall Street Journal* the previous August, reporting that up to 40 percent of his soybean fields were damaged by dicamba drift. According to reports of the trial, the two men didn't know each other. Wallace had called to discuss the source of dicamba injury on his cotton crop. The farmworker, Alan Curtis Jones, agreed to meet. Crops had been harvested, fields cleared, and the autumn air held the essence of the newly bare soil. Jones brought along his cousin and a gun.[22]

Monster pigweeds, illicit pesticides, battered crops, fist fights, murders. It all seemed too much. The killings, in particular, hung over every conversation at professional meetings. Everybody expected things to cool down in 2017, when the new low-volatile dicamba was widely available, fines for illegal applications were raised, and it looked like pigweeds had finally succumbed. Kevin Bradley, a professor at the University of Missouri, tallied reports of crops injured by dicamba around the country. In 2017 about 2,700 cases were under investigation and about 9 million hectares (3.6 million acres) of soybeans—worth hundreds of millions of dollars—injured by dicamba. By June the following year (2018), reports were already coming in of injury to cotton and soybean fields as well as vegetables, fruit crops, ornamentals, and hardwoods.[23] The seeds produced on soybean plants that had been damaged by dicamba grew into seedlings showing injury symptoms.[24] Without explanation, state agencies stopped reporting cases. Bradley and others who expressed concern, or suggested safety measures, were criticized or snubbed by the chemical industry.[25]

Some farmers refused to apply dicamba, even the low-volatile formulation. They didn't need it, couldn't afford it, or wouldn't risk damaging their neighbors' crops. They found other ways to deal with pigweeds. Still, sales of dicamba-safened seeds increased. They were more expensive than regular seeds, but farmers who didn't use dicamba bought safened seeds to protect their crops from volatile dicamba.[26] Even Mike Wallace's family, who continued farming after his death, bought the safened seeds to protect their crop, while refusing to use dicamba.[27] Industry touted the increased sales as proof of farmer satisfaction. Biotech seed company shareholders could not have been displeased.

Dicamba injury soon went beyond pigweed-filled crop fields. The damage shifted to other sensitive plants, including native species like redbuds, Kentucky coffee trees, and oaks. Street trees, landscape trees, and ornamental plants have shown distinctive cupped leaves, wrinkled stems, and contorted canopies.[28]

After three years of suppression by dicamba herbicide, the everlasting pigweeds reappeared. In fields where dicamba and 2,4-D had been used in continuous cropping, farmers started to see occasional pigweed plants that were not killed by the standard treatment. At first, farmers weren't sure what was happening. Researchers at Kansas State University confirmed that populations of Palmer amaranth evolved resistance to dicamba and 2,4-D.[29]

Selection pressure from repeated glyphosate and dicamba applications remains high in Kansas, across the Delta and the whole Midwest.[30] With the history of hybridization and pollen-dispersed resistance genes, there is nothing to stop dicamba and 2,4-D-resistant pigweeds from making their way across the continent. These herbicides, along with triazines, ALS, and glyphosate will soon be useless against pigweeds, even in fields of crops with stacked GMO genes. The trauma, anger, murders, and lawsuits will all be for naught.

Something to Tell Of

I looked out from my lab in Ohio toward the chaos in the middle of farm country and wondered where it was all heading. Maybe the response of pigweeds to biotechnology marked the end of the chemical weed control era that began with 2,4-D in the late 1940s.[31] Modern pigweeds had undermined attempts to use biotechnology to solve a really bad problem. The result was even more robust and widespread weeds, against which herbicides became increasingly useless. Farmers' experience with low-volatile formulations challenged assurances that toxins could be dispersed safely over thousands of acres. Dicamba volatility and mismanagement essentially forced farmers to buy the GMO seeds. This reinforced ethical concerns about intellectual property rights, labeling, corporate power, and monopolistic behavior.[32] And soon homeowners and foresters were tallying environmental damage to native species.[33]

On June 3, 2020, after farmers had invested hundreds of thousands of dollars on the high priced seeds and chemicals for the crops then in the field, the Ninth District Court vacated dicamba's registration, stating in part, "EPA entirely failed to acknowledge the risk that [over the top] dicamba use would tear the social fabric of farming communities. We therefore vacate the EPA's October 31, 2018, registration decision and the three registrations premised on that decision." With robust pigweeds resistant to triazines, ALS, and glyphosate, and dicamba out of the picture, maybe it was time to consider some of the nonherbicide options, as awkward and impractical as they once might have seemed. Studies in Georgia and Arkansas showed that cover crops and standard agronomic practices reduced the impact of pigweeds. That's what organic farmers were already doing.[34] Sandblasted coffee grounds, stale seedbeds, or even night tillage might yet have their field day.

But it won't be the end of pigweed evolution. Their ecological fitness is likely to be further enhanced by the latest efforts to use gene editing technology to kill weeds. A DNA processing system called CRISPR now allows scientists to make precise changes in the gene sequence of an organism.[35] CRISPR can cheaply delete, modify, or replace the letters of the genetic code. Applications of this technique for medical science to fix genetic flaws and alleviate human suffering are incredible.

But first it will likely be used to kill weeds. The technology is called a gene drive. It uses CRISPR to make specific changes in a weed's DNA, and to "drive" those altered genes throughout the population in a few plant generations. The system modifies the usual pattern of inheritance, so altered genes appear more frequently. A gene drive might disable the genes that make pigweed resistant to glyphosate. This would restore pigweeds' susceptibility to this herbicide. For modern dioecious pigweeds, a gene drive might target sex-specific genes to make either male or female plants less fit. The whole species would collapse.

Biotech labs around the world are dabbling with CRISPR gene editing technology to kill weeds. Pigweeds are a primary target. Exciting reports assert the potential for gene drives to wipe out Palmer amaranth.[36] Biotech companies are not waiting for ecologists or bioethicists to weigh in before pursuing these ideas.[37] When first deployed in agriculture, somebody will make a lot of money. And pigweeds will become even better known. For better or for worse.

Consequences of a gene-edited *Amaranthus* species are hard to predict. There are no good models of rapid genetic changes moving through an entire plant family. Edited genes could flow to nine hundred or so widely outcrossing *Amaranthus* species around the world, including some crops and others that provide ecological services in diverse ecosystems. Nobody knows how gene drives might function in a species with unusual ability to evolve mechanisms to resist herbicides.[38] And there's also the issue of Palmer's ability to produce genetically identical seeds through apomixis (without fertilization).[39] The offspring can have unbalanced chromosome numbers, and exhibit "curious tendencies to gigantism of seeds and other parts."[40]

All the pieces of this puzzle have not yet been assembled. But they offer the possibility that using a gene drive to wipe out modern dioecious outcrossing pigweeds might enhance weedy populations of mutant self-duplicating pigweeds with the potential for colossal growth. Robust pigweeds would be replaced by pigweeds of unmatched size and competitiveness. And they would already be resistant to most herbicides.

Now that is a scenario for ecological success that cannot be beat.

But it still could be better.

Suppose the CRISPR-edited genie will not go back in the bottle. It would not take the genius of Dr. Seuss to imagine such a scenario, that once released into the pigweed genome, there would be no way to stop the edited gene or to undo it. No way to put the brakes on if an altered gene spreads to useful species, collapses important ecological connections, or induces changes in other parts of the genome. That would alter ecosystems in unknown ways and make gene-driven colossal pigweeds even more successful.

To Make Your Heart Beat

The ongoing evolution of pigweeds raises questions about human-plant interactions that make for weediness. Out of seventy-five or so species of *Amaranthus* in North America, about a dozen became weeds in one situation or another. That's not a lot, but a higher proportion than in most plant families. Among all the species that have evolved herbicide resistance, the pigweeds have evolved more mechanisms to more herbicides than almost any others.[41] Thus, the ability to become a weed is not distributed evenly in the botanical world.

Palmer and waterhemp, in particular, are among thousands of species that were of no concern for the first twelve thousand or so years of agriculture. Modern practices, chemical farming, and sophisticated biotechnology put them—but not others—on the path to extraordinary weediness. In other words, the ability to evolve resistance is not the same in all species or families of plants.

The ability to evolve has evolved differently in some species than others. Pigweeds will likely continue to evolve no matter what technology humans throw their way—even night tillage, cover crops, microwaves, sandblasting and just anything attempted for pigweed control. Using them repeatedly over a large scale would impose agrestal selection pressure, but resistance is unlikely because of many genes involved.

Weeds—pigweeds in particular—continue to evolve in response to agrestal selection pressures. This truth has been obscured by fascination with profit-driven technology, especially biotechnology designed to undo the weedy consequences of the previous profit-driven technology. Human interactions with plants over thousands of years have not led to the acceptance of this reality in spite of our coevolutionary role in the creation and evolution of weeds. There's little indication that humans, for all our cleverness, might take a new perspective, change goals, appreciate connections, marvel at the life-sustaining gifts of the natural world, and find ways to live with that awareness—and with weeds—in different ways. So onward marches CRISPR, gene drives, and whatever follows. Pigweeds will have other genetic tricks up their weedy sleeves. Their ability to evolve multiple and novel responses to human manipulation is unlikely to fail the everlasting *Amaranthus* species.

Giant ragweed, *Ambrosia trifida*.

7

Ragweed

I wanted to have nothing to do with giant ragweed (*Ambrosia trifida*). And I wasn't especially enthralled with its sister species, common ragweed (*Ambrosia artimisiifolia*), though I knew I couldn't escape that one. But giant ragweed: no way.

That's how I started my job doing research on weedy plants in Ohio in the late 1980s. I'd been hired to work on problems farmers were having with weeds in reduced-tillage systems, like unexpected shifts to more troublesome species. Luckily, ragweeds, with their large blocky seeds, didn't seem too important in those systems. So I planned a series of experiments to explore changes among other species that infest fields, farms, and gardens. I was drawn to emerging ideas about species interactions, gene flow, and theories about population dynamics that didn't involve giant ragweed.

The professional reason: highly accomplished colleagues in Ohio and around the Midwest were already doing excellent work on giant ragweed.

The species had been widely researched and its biology was already well known. There was little I could add to the ongoing work.

The real reason: giant ragweeds are hairy and sticky and unpleasant to work with. They're coarse, tough-stemmed, and have a rank odor. And did I mention they are the main source of hay fever pollen for all of North America?[1] Two students had already quit due to allergic reactions to other weeds. The chance of finding a graduate student who wasn't toting a pack of antihistamines seemed remote. And besides, I'd heard that researchers in Tennessee were using chainsaws to remove giant ragweed from their research plots, hauling "truckloads and truckloads" from the field.[2] Who would want to work with a plant like that?

The research farm manager understood what was at stake. He was a stone-faced guy who likely slept in his plain brown hat. He had only one rule: no giant ragweed on the research farm. If giant ragweed got started, it would spread everywhere and contaminate the field research of six or eight other scientists. I embraced this rule. My experiments would be weedy, sometimes messy. My field plots would have many different species. But no giant ragweed in the mix. It would suppress all the others, ruin my research, and contaminate the farm for the next researchers who came along.

Maybe the Leftovers

The word *rag* has been linked to *weed* to disparage any number of plants. Here I use "ragweed" to aggregate two Midwestern species that present a scourge of weediness. I will focus mostly on giant ragweed, the more obnoxious one, and distinguish it when necessary from the insufferable common ragweed.

The two ragweeds are easy to tell apart. Giant ragweed can grow to heights over 5 meters (16 feet). It is one of the largest annual weeds, easily twice the height of wheat and soybean plants, an arm-stretch taller than corn.[3] The whole plant is rough, with leaves like large, three- sometimes five-fingered claws. Common ragweed, as the name implies, has spread more widely, its seeds infesting soils of most farm fields and gardens. It usually grows about knee high, though it can reach 2 meters (6.5 feet) with its lacey, fernlike leaves. Ragweeds are often confused with other plants, especially with goldenrod (*Solidago* spp), not because they look alike, but because they flower at the same time. When grey-green, nondescript ragweeds fill the air with sneeze-inducing pollen, the showy yellow-flowered goldenrod gets the blame.

Midwestern farmers often call giant ragweed "horseweed" due to its size and strength. It is also "blood-ragweed" after the red sap that drips from cut stems, or "crown-weed" from the seed shape. What North Americans call common ragweed is "short ragweed" and "Roman wormwood," a true misnomer elsewhere. The "rag" of both species has no clear origin—maybe the ragged edges of the leaves, maybe the handkerchief carried during fall allergy season.

The genus name is a mystery: *Ambrosia*, "Food for the gods." Good old Linnaeus gave the unattractive source of hay fever this lovely name. Nobody knows why. The connection between ragweeds and hay fever was not understood in his time. Maybe he had a sense of humor. Maybe he had in mind the pungent smell, one only the gods could love. Or maybe he just had a hankering for Aunt Jean's famous mélange of marshmallows, canned fruit, and Cool Whip.

Utilitarian, Like a Jeep

Ragweeds are in the Aster family. Unlike many Asters, evolution of ragweeds favored efficiency over beauty, functionality over frills. Daisylike petals are great for attracting pollinators; otherwise they're a waste of energy. Ragweeds would let wind, not insects, spread pollen to fertilize flowers. They evolved separate male and female flower structures on the same plant. Male flowers would concentrate on making pollen; female flowers would accommodate a single seed. The fruit would be simple, no parachutes, bristles, or hooks. Somehow, they'd find a way to get around. These rough, unattractive plants emerged in western North American gullies where steep slopes and valleys favored the evolution of fast-growing annuals.

Spread (or "dispersal") of ragweeds would depend on events that ecologists call a disturbance, something that disrupts an ecosystem. A tree falls in the forest, a fox digs a hole: soil is exposed, moved, and reshaped. Soil, nutrients, water, and microbes are vulnerable to loss.

This is where ragweeds come in. They are "pioneer" or "colonizing" species, the first plants to scavenge resources at disturbed sites.[4] The early colonizers send out roots that stabilize the soil and prevent erosion by wind or water. Those roots bring buried nutrients close to the surface where other organisms can access them. The canopy of ragweed leaves protects the soil from raindrops, intercepts sunlight, and uses CO_2 in photosynthesis. This brings solar energy into the soil in the form of complex carbohydrates. Leaves shade the soil surface, cool it, and provide sites where other plants, animals, and microbes can establish functional interactions.

All this sounds great. Compassionate even. Ecosystems depend on colonizing plants to protect and replenish disturbed sites. But colonizing ragweeds quickly dominate the plant community, produce most of the plant biomass, and suppress other species. Ragweeds behave as a "keystone" species by excluding slower-growing herbs, shrubs, and other flora.[5] They release allelopathic chemicals into the soil and suppress the growth of neighboring plants.[6] Eventually, ragweeds become victims of their own self-hating success and give way to other species. But when the site is again disrupted, the scruffy colonizers are among the first to appear.

Over thousands of years ragweeds hopscotched from one small disturbance to another. Their seeds are surrounded by corky fruit structures that allowed them to float toward the Mississippi watershed.[7] Birds and a few other animals helped as well.[8] The two ragweed species followed different routes. Common ragweed headed toward upland plains where it evolved greater tolerance to heat and drought. Giant ragweed did not spread as fast or as far, probably because it produces fewer seeds (100 to 300) than common ragweed (3,000 to 60,000). Moreover, giant ragweed seeds are much larger and about eight times heavier. The giant morph favored wet areas in more northerly climes, settling along river valleys and flood zones. When the glaciers receded, ragweeds occupied scattered patches among other native flora.

A Geek in the Woods

Common ragweed grew in the play area of our yard carved from abandoned farmland and exhausted soils. It stood at the edge of the sand box, in the garden, next to the swing set. I liked the tingling feeling of running the upper stalk between my fingers, popping off beadlike flowers. I tossed them into the air or against a wall to hear them bounce like soft gravel or threw them at a little brother. That's how grand life was when I was a boy in Ohio. I learned to recognize the seedling as soon as I was old enough to hold a hoe. My mother taught me how to spot it. A city girl transported to the country, she hadn't learned the names of many weeds, but ragweeds she knew. My sister was allergic to them, so we were supposed to yank out ragweeds wherever we saw them.

Giant ragweed I first encountered on the edge of a steep ravine where I hiked, learned the trees, and skipped rocks across pools of water. The woods was a different sort of play area, a place to explore what grew in the shade of tall trees. Water trickled in the creek surrounded by ferns and wildflowers. This is where I started to make sense of a world reflected in moving water, skittering dragonflies, and graceful trees. I pulled to my face the leaf of a large ash tree. Plants make the oxygen that I breathe; I exhale and feed the plants. Fields and forests are green.

I overheard my parents discussing an article in one of the gardening magazines. They were looking for unusual flowers—a pure white zinnia or a deep black rose. Whoever produced those would make a lot of money. I figured I had a chance. After all, my sister had ragweed allergy. With allergy season came allergy shots and my chance to pull the used syringes and needles out of the trash. I stored them in an old cigar box. Now with some ink and paint and maybe a little gasoline or furniture polish I could start injecting plants and watch the flowers turn color. As I hiked the woods, I fantasized about the success of my experiments.

Oversharing, Perhaps

Ragweed allergies begin with the male ragweed flowers. They are stacked at the top and ends of branches where rows of inverted (down-facing) flower heads hang along spikelike stems. Imagine tiny sunflower heads, diameters less than a pencil, hanging upside down. Each head holds a cluster of ten to fifty individual florets that open and release pollen at different times. What passes for petals are five green flaps. Inside are little stalks (stamens) tipped by cone-shaped pollen-producing anthers. Attached to each anther is a thin, flexible claw.

Now look closer. Inside the circle of anthers and claws is the pistillodium, an appendage unique to ragweeds. When the mood strikes and a flower opens, the claw straightens. It ruptures the anthers and exposes the pollen. Meanwhile, the pistillodium elongates rapidly, pushes past anthers and claws, and thrusts a wad of pollen into the air. Hairs surrounding the tip of the erect pistillodium sweep out any remaining pollen as the stalk shrinks back into the flower.[9] You can't fault them for lack of effort: a single ragweed plant can release up to 10 million pollen grains per day.[10]

The female flowers are discreet. They rest in the axils of upper leaves and branches, with two antlerlike stigmas whose short, curved hairs catch wind-borne grains of pollen. The flower is simple, unadorned, designed to birth a single seed.

One of the keys to ragweed's success and allergenicity is found in and on the outer wall of its pollen. A pollen grain is a space capsule that carries male germ cells on their journey to combine with female gametes and make a new plant. It would make sense if pollen jerked into the air by male flowers simply fell on the female flowers waiting below. But these females rarely allow self-pollination. When they sense pollen from males of the same plant, female flowers block pollen tube penetration and self-fertilization. This seems inefficient, but it ensures the mixing of genes through cross-pollination.

With a stiff wind, ragweed pollen can travel 1,000 kilometers (620 miles) or more.[11] So plants in one habitat can mate with plants far away in different environments. This way, genes for many adaptive traits, like drought tolerance and herbicide resistance, can be spread through the ragweed metapopulation at a global scale.[12]

The wall of a pollen grain is covered with scores of sharp-pointed spines designed to attach to the stigma of a female flower. The spines also can attach to the lining of the organs and tissues that help you breathe. Female flowers recognize the proteins in the pollen wall and either reject them, or accept them to make a seed. The same proteins are recognized by human mucosal cells, which either ignore them, or are irritated by them to make a sneeze. Pollen wall proteins—not the germ cells inside—are the cause of hay fever. So pollen grains with intact walls remain allergenic even after germ cells die. When the immune system identifies the pollen as a foreign object, the body responds with an attack that includes the release of histamines. Then come the watery eyes, runny nose, wheezing, itching and general malaise.[13] For some, this allergic reaction can trigger an asthma attack as well.

Workin' Over Time

I worked hard to not work with giant ragweed. My research plots would be confined to part of the farm I could manage by myself. Unfortunately, the fields I was assigned had been well maintained and didn't have enough weeds to my liking. So I devised a plan to increase the diversity of weed species. At the end of the growing season, I contacted grain elevators around the state, asking if they would give me a few sacks of the weed seeds they screened from harvested soybeans. It was a good way to get to know the grain dealers. Most had never met somebody willing to drive across the state to collect refuse weed seeds. I targeted areas where farms had weeds that interested me, like velvetleaf, pigweeds, lambsquarters, foxtails, and others. I made it clear: no giant ragweed.

This is how I first met Dwayne, a tall, serious farmer with tousled black hair and a scraggly beard. I ran into him at the local grain elevator where I first collected weed seeds. When he learned what we were up to, Dwayne pointed us toward other elevators to visit and told us which to avoid if we didn't want ragweeds. He helped me stack the pickup with burlap bags full of weed seeds while I explained my plan to scatter them over the fields using a fertilizer spreader. I would let those seeds germinate and grow to produce more seeds, so over time, the fields would have a seedbank that represented the whole weed flora of Ohio, save one obnoxious species.

Dwayne just listened. He ran a grain and hog operation north of town. If I wanted to work on real weeds I should stop by.

The Flow of Progress

Ragweeds and humans have been entangled in each other's evolutionary development for a long time. At least five thousand years ago, people in the eastern Mississippi River basin began moving out of caves to settle near waterways and flood plains. When they started to grow food, ragweeds were already there. Wild sunflowers and other plants provided high nutrition seeds. People selected the most favored ones, and over hundreds of years domesticated sunflowers as a crop. Ragweed plants grew there as well, but it's uncertain whether anybody tried to domesticate them.

Among the people who study such questions is Dr. Kristen Gremillion, a colleague in the Department of Anthropology. She has an extraordinary academic record based in part on—and I say this with the utmost respect for her expertise and accomplishments—many hours picking through ancient human feces. I've been fortunate to interact with Professor Gremillion because she is an authority on the origins of agriculture in eastern North America, and especially the evidence—such as it is—for what was actually in the diet of early inhabitants.

Data that she and her colleagues strained to excrete from cave dwellings in Kentucky showed that around three thousand years ago, humans used local plants to supplement hunting and gathering. Evidence that ragweed might have had a role as food, according to Dr. Gremillion, comes from a single cave sample of paleofeces (yes, that's what they call it) containing small seed coat fragments. Was it eaten on purpose? Was it gathered accidentally along with sunflowers? Nobody knows. Without further evidence of widespread enjoyment of ragweed granola, it's likely that from the earliest days of agriculture in North America, ragweed, in spite of its large energy-packed seeds, was—well—just a weed.[14]

Ragweeds moved with early farmers who used fire to manage weeds in mound planting systems. The seeds were collected for ceremonial, medicinal, and other practical uses.[15] Over time, indigenous farmers expanded their fields beyond streambanks to upland sites cleared for maize, beans, and squash. The newly disturbed sites were ideal for plant colonization. Common ragweed moved into settlements and towns, carried unconsciously in crop seeds and by animals. By the time massive inflows of immigrants changed the landscape forever, common ragweed had reached across the continent and into gardens and crop fields. Giant ragweed moved slowly east and north, still hovering along waterways, roadsides, and field edges.

Opportunistic pioneers and hopeful settlers pushed the indigenous people aside to manifest the destiny of ragweeds. Large areas of forest were chopped and burned, setting the stage for advances in ragweed expansion. Cleared ground made way for towns, crops, and colonizing ragweeds. A few people, including Benjamin Franklin and botanist John Bartram, made note of changes to the environment where trees were removed. Captain Edward Johnson, the first historian of colonial New England, stuck thermometers in the soil and reported higher temperatures where forests were cleared.[16] To American entrepreneurs, the settlers were making improvements on nature, transforming the wasteland into a garden while civilization marched across the landscape.[17]

As the agrarian economy began to bloom, ragweeds gained a reputation. Writing in the 1840s, William Darlington's first book on American weeds described common ragweed as a "worthless weed [that] occurs in most cultivated grounds . . . always ready . . . to make its appearance." Giant ragweed was a "coarse ugly weed . . . sufficiently common, and worthless, to . . . every farmer who desires to keep his premises clear of such nuisances."[18] Although worthless, neither species was important enough to be included on Darlington's list of "pernicious and troublesome weeds."

Clearing eastern forests for settlement and farming hadn't taken very long. But opening the whole Midwest meant clearing and draining extensive wetlands, habitat for hundreds of species of water-loving flora and fauna. Giant ragweed thrived at edges of swamps and on hummocks in the middle of wetlands. Pioneer farmers dug ditches and cleared streams to make land farmable. Tubular clay tiles became available around 1860, and in ten years, five thousand miles of drainage pipes made Ohio's ragweed-infested Black Swamp the flattest and most productive farmland in the region. Vast wet prairies were opened to agriculture in Illinois, Iowa, and Minnesota. By the end of the drainage boom, about a quarter of the soils from western Ohio through Iowa had been tile drained. Transforming the landscape from fur, fowl, and fen to corn, soybeans, and potatoes destroyed about half the nation's wetland, 5 percent of native Midwestern flora, and most elm and ash swamp forests. This unleashed the weedy potential of giant ragweed on some of the richest agricultural land in the country.[19]

By the end of the 1800s, ragweeds could be found in their favored habitats throughout North America. They followed in the footsteps of progress, the first to colonize urban areas, roadsides, trash heaps, riverbanks, and railways. As much as the bald eagle, the ragweed became an apt symbol of America—the national weed perhaps.

Menace of Distinction

As ragweed populations expanded, so did the pollen cloud over the continent. The distinctive spines on ragweed pollen grains make them easy to identify in soil core samples extracted from the earth. This pollen can survive intact at great depths for tens of millions of years and is used as a marker of geological time.[20] Geological, environmental, and pollen data help trace evolutionary patterns in plants along with climatic conditions over eons. Geologists and palynologists (pollen experts) report that the total amount of plant pollen in the atmosphere over North America has fluctuated little over most of the last sixty thousand years. But beginning around the late 1700s, there began a huge increase in the ragweed component of the pollen cloud.[21] This coincides with large-scale forest clearing, prairie turning, and wetland drainage for agriculture and urban settlement. It also coincides with a rise in atmospheric CO_2 from about 250 to over 400 parts per million (ppm).

The overwhelming majority of scientists, historians, and allergists hold the view that this rapid increase in atmospheric ragweed pollen is related to increased human activity. The veracity of this view means nothing to hay fever deniers, who have been around since the mid-1800s, well before the disease was attributed to ragweed pollen. After all, pollen counts go up and down; people have experienced hay fever and asthma for centuries.

Hay fever was first reported as a disease in 1819 by British physician John Bostock. He described his own case but did not assign a cause. Dr. Morrill Wyman, an American physician and hay fever sufferer, connected the ailment with ragweeds, but not to pollen. He carried a bag of ragweed plants to the White Mountains of New Hampshire, a fresh-air haven for "hay feverites." Wyman opened the bag in the faces of hay fever sufferers and recorded sneezing, itching, and runny noses. Such was the nature of informed consent in the 1800s. Nevertheless, he attributed the response to the plant's aroma, making no mention of pollen.

Charles Blackley, an English homeopathic physician, likewise suffered. He applied dust samples to his tongue, lips, skin, and nostrils, and found that hay fever symptoms were elicited by those containing pollen. In a classic description of a skin test, Blackley evaluated 35 types of pollen, mostly grasses and common English weeds. He attributed the disease to pollen, but ragweed was not yet common in England, so neither species was among those he tested.[22]

Blackley's pollen theory was discounted by Bostock, Wyman, and others. For a different approach, the American physician George M. Beard conducted a survey among hay fever sufferers who spent summers at mountain and seaside resorts. Beard deduced that the disease was restricted to the educated wealthy class. Neither ragweed nor pollen had anything to do with it. Hay fever was "excited" by "American nervousness." It was akin to certain types of hysteria, insanity, and drunkenness. The underlying cause was the "very rapid increase of nervousness in modern civilization . . . distinguished . . . by these five characteristics: steampower, the periodical press, the telegraph, the sciences, and the mental activity of women." Also contributing were unspecified "personal habits of indulgence."[23]

At the end of the American Civil War, it fell to Dr. Elias J. Marsh to carry Grant's cease-fire orders to units still fighting. Later, Dr. Marsh saw hay fever in many of his patients and repeated some of Blackley's skin-test experiments. He concluded that hay fever was caused by pollen in the atmosphere, which irritated the respiratory system of susceptible people of all classes. In spite of his status as medical director at West Point and reputation as the "man who stopped the war," Marsh's conclusions were dismissed. Sir Morell Mackenzie, British physician, knighted Grand Commander of the Royal House Order of Hohenzollern, also repeated work of Blackley. He established that while "pollen is the essential factor" in hay fever, the disease appeared only among the upper class in England and America due to their racial superiority.[24]

By the late 1800s, hay fever became a mark of social distinction. Come August, thousands escaped to the White Mountains, New York's Adirondacks, and Colorado's plateau seeking respite. Bethlehem, New Hampshire, had been a dreary town until it was transformed to meet the needs of asthmatic East Coast intellectuals. Nine hotels sprung up around Mount Washington, some charging more than hotels in New York City. American and Canadian Hay Fever associations were established as exclusive social clubs featuring witty entertainers and jocular, puffy-eyed humor. Famous tourists included Daniel Webster, Ralph Waldo Emerson, Nathaniel Hawthorne, and Henry David Thoreau. They came to escape the pollen cloud, discuss great ideas, watch birds, enjoy delicious food, and listen to beautiful music. The luxury of the surroundings and wildness of the forests provided a sublime experience of nature as it was intended, a verdant contrast to corrupt urban life and toil, a poultice to the maladies of civilization.[25]

Hay fever associations engaged with local towns to purify their environment by encouraging the cutting of weeds that might cause irritation. They proposed bans on growing corn and vegetables (and their weeds) near the resorts to maximize the separation from agricultural contamination of the natural world.

Hay fever deniers were not deterred. Ordinary town folks suspected the hotel owners of manufacturing and promoting the disease among a privileged class of sniffling snobs. The fact that hay feverites formed societies and publicly paraded their association with the ailment and its glamorous sufferers reinforced a conviction that the so-called disease was a Gilded Age display of conspicuous consumption and cronyism. White Mountain locals accepted no connection between asthma and atmospheric pollen load or human-mediated environmental disturbance. So-called hay fever was a so-called disease of weak, self-absorbed, and privileged prigs who expected special treatment and consideration.[26]

And then in 1878, the inevitable happened: ragweed was spotted along railroad tracks near Littleton, a village just north of Bethlehem.[27] The notion that farmers were responsible for ragweed had led everyone else to ignore the menace in their midst. The fight for extermination was on. Few people could distinguish ragweed from botanical imposters. Every green plant with ragged, frilly, deeply divided leaves became suspect. With this wider conception of what constituted a "ragweed," the cutting, hoeing, and pulling efforts expanded to every ditch-bank, refuse pile, abandoned lot, and wayside.

Few things advance the ecological success of a weed beyond an organized attempt to exterminate it. Every removal of ragwort, yarrow, marigold, and half a dozen other misidentified species created a small disturbance, a potential site for establishment of ragweed. The pollen pusher spread at a pace that some considered inconsistent with natural processes, driven by incongruous forces of modernity. This was not just a homely weed of farm fields; it was an environmental scourge, a public health nuisance, and a personal affront to those cursed by hay fever.

Local hay fever associations around the continent knew their enemy. Never mind that most didn't know ragweeds from garden flowers. Pollen allergies affected about 10 percent of the general population and about 100 percent of Hay Fever Association members. There was no cure, and many lives were made miserable by it. Association members, with their social and political connections, organized efforts to pass ordinances prohibiting all weed growth and launched campaigns to eliminate ragweeds specifically. Departments of public health in cities around North America developed eradication campaigns along roadsides, waste areas, railways, and urban construction sites.

Community groups joined the cutting, mowing, and rogueing. All sorts of salts and petroleum products were dumped onto soil to kill ragweeds. By the 1930s, the marvels of organic chemistry reacted. An article in *Science* described efforts to find practical uses for a new chemical called 2,4-D. Studies suggested it might halt ragweed's pollen production. Both species were quite sensitive to this chemical. Sprayed with 2,4-D, ragweeds twisted and curled; their flowers wrinkled, refused to open, and released no pollen. The way to solve the health crisis was to spray 2,4-D out the back of a truck, shooting fifteen to thirty meters (fifty-one hundred feet) into the roadside. Next came a fog sprayer spewing oil mixed with 2,4-D. Ragweed eradication and pollen-free air were just a few thousand-gallon spray tanks away.[28]

A model ragweed education program was developed in Hartford, Connecticut. The health department made posters to display at health fairs and exhibits throughout the city. There were 4 posters on communicable diseases, 5 on syphilis, 2 on dental care, 80 on cancer, and 210 on ragweed eradication.[29] Similar programs began in Chicago, New York, and other eastern cities. New York's "Operation Ragweed" began in 1946 and ran for nine years. They cut the city's ragweed population in half. A single hectare of ragweed can release sixty-six kilograms of pollen (about fifty-eight pounds per acre) that floats hundreds of miles in the air.[30] The operation produced "no definite tendency for the pollen index of New York City or any other city to decline."[31] Nevertheless, hay fever awareness had been raised, and many people, like my mother, learned how to identify ragweeds.

Engines of Polemochorites

These homegrown efforts to eliminate ragweed would ensure its success across North America. But to colonize lands beyond North America, ragweeds needed a path to planetary pestilence that doesn't fit the pattern of most weeds. The seeds float, but they're big and clunky and not easily transported. The little projections on top that make them resemble a crown are not designed to attach to fur, skin, or human contrivances. The seeds have no parachutelike pappus, spines, sticky hairs, or dispersal appendages. Ragweed—seeds, stems, leaves—have no commercial value, role in trade, or place in industry. Global domination would require international coevolutionary partners in a relationship born of something besides attractiveness, usefulness, or desire.

Global ragweed dispersal has mostly been unintentional and cluelessly tangled with conflict and commerce. It started in the 1700s with maritime trade via shallow, top-heavy, unsteady, wooden vessels. To increase stability, ships took on loads of heavy materials or ballast. Rocks, sand, and soil were shoveled into cargo holds at one port and dumped out on another continent. Along with ballast came all manner of plants, microbes, insects, and other organisms. This was the source of ragweed first reported in Europe, near waterways where ballast was jettisoned during a ship's approach to harbor.[32] This was the most likely source of ragweed introductions to Germany (1860) and France (1863).[33] The oldest record from Asia is also a portside observation, from Japan in the late 1800s.[34] Such alien plants around shipping lanes and ports were curiosities among local botanizers who walked the shorelines and quays hoping to sight a species new to them. No one would have seen them as plants of significance.

The world was ultimately made safe for ragweeds when America ended its isolationist leanings during the First World War. The nation first projected its power across the Atlantic as one would expect, with horse fodder. European armies of 1914 still depended on thousands of horses for troop movements and on the battlefield. France looked to the United States for help. President and hay fever sufferer Woodrow Wilson eagerly obliged. Hundreds of American horses were drafted and shipped to Atlantic harbors. With them went American hay, about twenty kilos (forty-four pounds) per horse daily. In the hay were American ragweeds and their seeds. American soldiers followed into France in 1917. They, too, carried ragweed seeds in and on clothing and machinery. At ports of landing and wherever troops were stationed, small populations of the raggedy alien became established.[35]

Nothing like a good war to trigger environmental devastation and create ragweed habitat on a massive scale. Troops gotta move. Trees, farms, and houses can't get in the way of trucks, artillery, rocket launchers, and mammoth tanks. Trenches must be dug. Fields must be cleared to stockpile fuel, oil, armor, munitions, and poisons. Land will be burned, bombed, or pushed aside. Into this disturbance came the botanical colonizer, its seeds carried by soldiers and machines. They thrived on the spoils of fuel spills, unspent ordnance, chemical dumps, pesticides, hazardous waste, and rubble. Ragweed genotypes that tolerated the toxic conditions filled the air with pollen and passed their genes on through generations, a contorted, perverse version of agrestal selection.

Ragweeds became archetypical polemochorus plants, those dispersed through armed conflict.[36] They followed American troops throughout Europe, where small ragweed colonies grew into large populations. Later, in occupied Japan, common ragweed settled in urban areas while giant ragweed colonized marginal areas throughout the islands. The colonizer's seeds moved next to Korea on the boots of American soldiers in the early 1950s. Still today, ragweeds remain heavily protected at abandoned military camps along the 250-kilometer (160-mile) DMZ, a polemochorus testimony to human tragedy and weedy success.[37]

Health consequences of an atmosphere increasingly filled with irritating pollen were a small part of war's aftermath. Europeans and Asians had little exposure to ragweed pollen or associated hay fever. Other sources of asthma were well known, but ragweed pollinosis was new, and it took several years to figure out the source of the new malaise and irritation.[38]

Ragweeds prospered from historical agricultural policies of the USSR. Joseph Stalin aimed to prove the superiority of communism by producing wheat on a massive scale. First, he had to "liquidate" peasants and force them into collective farms. Thousands of farmers protested by slaughtering livestock and destroying farm equipment. Over 1.5 million Ukrainians refused collectivization. Eventually, the army successfully collectivized the peasants, excepting the few million killed or in prison camps. Without functional machinery, horses, or cattle, food production collapsed. Famine killed another few million in the countryside.[39] Human suffering and deprivation were ragweeds' opportunity. They colonized degraded and disturbed fields and flourished across the North Caucasus and Southern Russia. By 1940, they were spewing out pollen from neglected and poorly managed land from Ukraine to Austria and throughout Hungary, where ragweed became known as "Stalin weed."[40]

Cold War reconstruction drew ragweeds further into Europe and parts of western Russia. When socialist collectives in the former USSR were finally dismantled, farmland was put in the hands of people without capital or knowledge of farming. Ragweeds colonized fields that were subsequently abandoned, left uncultivated, or farmed poorly.[41]

When hell erupted again during wars in former Yugoslavia in the 1990s, soldiers and machinery again facilitated ragweed dispersal. Seeds exploited misery across rural and urban areas converted to wasteland in the wake of bombing. Ragweeds fed on the chaos of fields abandoned when farmers became refugees, soldiers, or died in the fighting.[42] They colonized land abandoned by population displacement, and land that had been converted to minefields. Twenty years later the landmines were still there. So were the ragweeds.[43]

Conspicuous Consumption

In the early twentieth century, the United States became a major exporter of grains around the world. Until this time, technologies for separating contaminants from grain were rudimentary. Bags of farm commodities shipped from US harbors were haphazardly contaminated with weed seeds from across the country. To standardize grain quality, export rules were implemented beginning in the 1930s to restrict contaminants, or "foreign material." The limits on foreign material were set at 1 percent by weight; 5 percent for lower grades. In 1940 alone, the United States exported 30 million metric tons (33 million US tons) of wheat to Europe. If only one-tenth of 1 percent of wheat exports had 1 percent weeds, about 5 billion giant ragweed seeds could have accompanied clover, corn, soybean, and sunflower exports. Farmers across Europe and Asia sowed them in well prepared, fertilized, and tended fields. The seeds were fed to, and passed through, farm animals that roamed pastures and fields. Chickens, hogs, and cattle spread and fertilized them.

No wonder ragweed was already part of the resident flora when global economic expansion began in the 1950s. Both species colonized new construction sites, roads, railways, and canal networks. Infestations crept along river systems, into urban zones, and mountain island habitats.[44] Changes in flora and fauna accompanying ragweed's appearance were slow and imperceptible steps in homogenization of global ecosystems. In a few years it would be as common to see ragweeds in Europe and China as garlic mustard in Illinois, *Ailanthus* in New York, KFC in Shanghai, and McDonalds on the Champs-Élysées.

Inevitably, the entertainment industry could not resist a bumbling bully. From the Po Valley to the Apian Way, ragweed toured with the Italian circus—acrobats, jugglers, clowns, dancing bears, the whole bit. They trekked from town to town, dragging animals, their feed and fodder. Circus tents appeared on the outskirts, where large talented animals grazed, scratched the soil, and deposited ragweed seeds along with everything else. In this way, ragweeds stretched across the countryside at about six to ten kilometers (about four to six miles) per year.[45]

Giants in the Earth

As randomly introduced and dispersed plants, the two ragweed species had equal opportunity to colonize disturbed sites in Europe and Asia. But even during the first few decades of American chemical agriculture, giant ragweed was still considered a weed of stream banks and waste areas, whereas common ragweed was widely distributed. Where I grew up in Ohio, a giant ragweed sighting outside of a roadside ditch was a noteworthy event. As soybean acreage increased so did the reliance on herbicides as the primary means of weed control. Farmers were encouraged to tear down fencerows and drag crop planters right through those inconvenient ephemeral streams and waterways. Giant ragweed habitat became soybean and corn habitat. Giant ragweed seeds got caught in field equipment and dragged to upland sites. Harvester combines gathered weed seeds and scattered them across fields along with crop residues. By the early 1980s, giant ragweed completed its journey from riparian and field boundary areas into Midwestern farmland. Genotypes adapted to agricultural habitats became distinct from those of riparian areas.[46]

New conservation tillage practices allowed farmers to plant major grain crops with less soil tillage. Without tillage to bury them, giant ragweed seeds were too large to bury themselves in the compacted soil surface. Exposed seeds could not survive, and farmers delighted that giant ragweed might decline in fields with reduced tillage.

But giant ragweed found ways to butt into crop fields with or without tillage. It developed a peculiar relationship with the humble earthworm (*Lumbricus terrestris*), which European settlers had introduced to North America. Earthworms collect giant ragweed seeds from the soil surface and drag them into protective vertical burrows. Why the worms do this remains unclear. In the burrows, the seeds are safe from predators and buried at optimal depths for germination. Nutrients and moisture in the burrows help ragweeds get a good start. Since many ragweed seeds are packed together in the burrows, there is intense competition among the emerging seedlings. Only the most robust genotypes survive.[47]

In intensively managed crop fields, farmers add fertilizers to soils to produce high yields. Giant ragweed plants are especially well adapted to accumulate more soil nitrogen than they need for growth. The bloated ragweeds produce roots profusely, which allows them to take up more of other soil elements, like phosphorus and potassium. The result is earthworm protected and farmer fertilized ragweeds of enormous size and growth potential. Plant scientists call this kind of self-indulgence "luxury consumption."

Weed for a Lifetime

On a spring morning in the mid-1990s I got a message from Dwayne asking if I might drive by and check out a weed near his hog barn. I had become familiar with his fastidious management. He had been able to keep "giant rag" out of his fields. He talked about his vigilance all the time. He took notes and knew where every weed patch was on his farm and his neighbors' as well. He watched how the neighbors worked to control it.

Their crops were sometimes choked with ragweeds taller than the corn. At harvest, they simply drove the combine around the ragweed patches. To avoid giant rag, Dwayne wouldn't buy hay harvested from neighbors' fields, borrow their equipment, or drive his tractor down his neighbor's lane. On summer evenings he walked his fields until dark, hoe in hand, ready to chop out anything that remotely resembled giant ragweed.

I parked at the field edge and walked a dirt path toward the pungent commotion of the pig barn. A few paces away, among freshly planted corn, was a patch of bright green seedlings with large fleshy cotyledons and clawlike leaves. Dwayne would not be happy to know giant ragweeds were streaked across the field like a rough brush stroke. Porcine grunts accompanied my picture taking until Dwayne showed up in his blue pickup.

He couldn't explain it. He had done everything possible to keep it out of his fields. He paced back and forth, rambling theories of how it happened. Maybe it was that gulley-washer a few weeks back; maybe rag seeds flowed off the neighbor's field along with the water and silt. The water hadn't run that close to his barn, but the seeds surely floated onto the farm. Now he had to figure out what to do about them. If he didn't knock them back this year, he'd fight them for the rest of his life. And he knew the neighbors used the same herbicide so many years in a row that the ragweeds were likely resistant.

Back at my office, I got a call from somebody in the veterinary school. They were doing a study on roadkill. And I thought *I* had problems explaining my job to my children. Anyway, would I be willing to look through the stomach contents of a few animals to help identify what they were eating? It wasn't ancient feces, but still I couldn't refuse. The next day, a stack of plastic petri dishes showed up in the lab, clean and surprisingly not disgusting. There was stuff that looked like gravel, bits of stems, and seeds of a few native grasses. Three samples—from a Canada goose, a wild turkey, and a groundhog—had giant ragweed seeds. The goose had 19 seeds in its crop. They wouldn't let me keep the seeds to test their germination, an area of study begging to be explored.

I told Dwayne about this next time I visited the hog barn field. He was still trying to figure out how giant rag had weaseled its way into his life. His eyes lit up at the suggestion of Canada geese. He'd seen a big flock land after plowing his fields. They nested along stream banks on the neighbor's farm near a patch of giant ragweed. That had to be the source.

Not much you can do about Canada geese, moving in with urban sprawl creeping toward his farm.

But soon Dwayne wasn't too sure. It didn't make sense that geese would go to just one side of the field. Anyway, he had run the cultivator through the ragweeds a couple more times. Escaped plants were already halfway to his knees. He'd spray something on them if he felt he had to.

I was pretty sure he'd eventually feel he had to. But what to spray? Ragweeds had left farmers without good options. The seedlings emerged early in the spring. Early plowing or herbicide spraying only destroyed part of the population. Another cohort of seedlings emerged with the crop or later in the season. Herbicides farmers used were good, but not great. Many farmers simply sprayed more, ultimately exerting severe selection pressure for tolerant or resistant biotypes.

A few days later I ran into Dwayne again, in the grocery store. He'd finished planting his soybean fields and there was a new litter of piglets. I should stop by to see them. We talked about the rain and low corn prices. Then he glanced both ways and spoke quietly. He had figured it out about the ragweed. Last fall the grain bin was getting low and feed prices were high. Somebody was selling corn cheap. It would fill in as hog feed until he harvested his own corn. The grain looked a little rough, with stems and cob pieces. But the hogs had to be fed. It didn't make sense to run cracked corn through a cleaner; it was hog feed, not seed corn. Hogs will eat anything. You can figure out the rest. He spread the hog manure near the barn. Nothing ragweeds like more than hog manure. He looked around again. His voice tightened, "And now I've got the biggest, greenest, fastest growing goddam ragweed in the whole county."

We paused in front of the frosted flakes. After a spell, he had more to say. He told me the neighbor had sprayed glyphosate on some giant rag. Something strange was going on. Leaves on the sprayed plants turned brown and died quickly. Then the plants started to grow all over again from buds low on the stem.[48] "That weed plays with your mind, man," he said. I had read about this rapid desiccation response, unique to ragweeds. I didn't tell Dwayne that as an obligate outcrossing species, the genes for that response spread through pollen. He already knew that nothing pumps pollen into the air more imposingly than a ragweed.

I tried to change the subject. I reminded him how we met at the grain elevator and how I appreciated his help. The weed seed collections for my experiments had worked great. Within two years I had a diverse mixture of weeds. I had some interesting research results. I'd share the data with him.

Then it was my turn to pause, glance both ways, and lower my voice. At the end of the second year I found a couple of giant ragweed seedlings in one of the plots. I didn't want to yank them out and mess up the experiment. But I couldn't leave them in the field and mess up the farm. I decided to leave them and collect their seeds so they wouldn't spread. But it didn't matter. The next year the whole field erupted in giant ragweed. I had polluted the whole farm with giant rag. The farm manager was pissed. Worse, now I couldn't escape doing research on ragweeds.

It was the only time Dwayne cracked a smile. "Well, looks like you and me are in the same business. If you need a place to do some real rag research, come by the hog barn."

Ragweed Meltdown

In 2013 I was able to attend the International Weed Science Society meeting in Hangzhou, a beautiful city in China's northwestern Zhejiang Province. When I finally slept off the time warp and travel disorientation, I looked down from the window of my hotel room and into a new construction site. I thought about Dwayne staring into his corn field with some kind of knowing dismay: is that really giant ragweed? Why am I surprised?

Ragweeds were among the flowers to bloom in China's transformation to economic power. There is little information about ragweeds in China from their introduction in the 1930s until the 1980s. During the intervening fifty years *Ambrosia* colonialists made a great leap forward in dispersal and annoyance. Economic progress created an ecosystem disturbance of phenomenal scale and impact. Polluted soil and water and lost native biodiversity have provided nearly endless opportunities for ragweed exploitation.[49]

With China's new openness, millions of tons of US grain have been imported, and the issue of grain quality has resurfaced. China imports mostly lower grades. Officials have complained that too many weed seeds—ragweeds specifically—and herbicide resistance genes, were shipped in US soybeans.[50] The seeds were also imported as contaminants in "renewal resources" such as industrial wastes and scrap materials.[51]

I should not have been surprised to see giant ragweed in Hangzhou and along the train tracks to Guangzhou. China's American ragweeds flourish on industrial pollution. They dominate soils contaminated with all sorts of toxic chemicals, heavy metals like cadmium, arsenic, and lead. About 20 percent of arable land in China, an area about the size of Kansas, has excessive levels of metals, mostly near industrial coastal areas, some around iPhone and Nike factories.[52] Ragweeds thrive on these sites because they are "hyperaccumulators" of heavy metals. Engineering firms have encouraged planting ragweeds to clean up contaminated soils.[53] More intriguing is harvesting giant ragweed to make a charcoal-like product (biochar) to remove carcinogenic explosives (TNT and RDX) from groundwater.[54] Plans for future economic development guarantee additional land and water contamination by mining, dumping, dredging, smelting, fuel refining, and other hallmarks of progress across the landscape.

As ragweeds traveled through Asia, as in Europe, their evolutionary adaptation has increased. At home in North America they remain mostly weeds of agriculture and disturbed urban environments. But elsewhere in the world, they also behave as invasive species in less intensively managed sites such as forests and parks. They blur the distinction between "weed"— a phenomenon of managed environments—and "invasive plant"—which persists without human intervention. Ragweeds are among the few such crossover species. This is aided by ragweed's high genetic variability and plasticity. These give ragweeds a wide niche breadth, or ability to make use of a broad range of resources.

Among the adaptations in European and Asian environments are epigenetic changes. By escaping their homeland, ragweeds escaped their enemies and pathogens, and no longer have to defend themselves from predation. Resources that were needed for self-defense in their homeland are put into growth, competition, and reproduction in the new habitat.[55] They exhibit evolved increased competitive ability (EICA). These ragweeds have more, larger seeds that germinate faster under a wider range of conditions. The seedlings tolerate frost better.[56]

After colonizing sites to monopolize resources, EICA ragweeds increase the damage caused by other invasive species. They stick around longer at a site before yielding to succeeding plants. Their greediness discourages pollinators and suppresses species that depend on pollinators. This makes them part of an invasional meltdown, a chain reaction of disturbance that favors further ragweed establishment and growth.[57] They've become part of a homogenized global ecosystem where the same mix of invasive species occupies habitats around the world, blurring the distinctiveness of ecosystems once considered unique and natural.

Wheezing into the Anthropocene

The earth and its inhabitants are now firmly in the Anthropocene, a geological epoch marked by human impacts at a planetary scale. All of earth's interacting biological, geological, and chemical systems are now shaped by anthropogenic (human-mediated) activity.[58] For some people, putting humans in the driver's seat of essential planetary processes might raise little concern. Yet the way humans have driven processes on smaller scales might give some clue as to how things are likely to proceed.

Ragweeds are among the plants that benefit from climate change. Look inside a ragweed leaf where CO_2 is absorbed for photosynthesis. The main enzyme of photosynthesis, "Rubisco," captures either CO_2 (carbon dioxide) or O_2 (oxygen). Captured CO_2 is used to make sugars needed for plant growth. But captured O_2 steals and wastes CO_2 in a process called photorespiration. These are called "C3" plants because the first stable product of photosynthesis has three carbons. The C3 plants like ragweeds operate more efficiently at higher levels of CO_2. As atmospheric CO_2 concentrations increase, ragweeds make more ragweed biomass—taller stems, larger leaves, more seeds, and more pollen.[59]

Moreover, climate change increases ragweed pollen and its allergenicity. As a graduate student in 1980, I calibrated the photosynthesis equipment at about 312 ppm. This morning it was 414 ppm, a 33 percent jump, enough to increase average daily pollen over 230 percent.[60] With higher global temperatures, ragweeds germinate earlier, flower longer, and grow later into autumn.[61] The pollen spewing season has increased from thirteen to twenty-seven days over fifteen years as global temperatures have risen about 0.15 C.[62] In cities, high temperatures and CO_2 interact with diesel exhaust and ozone to increase ragweed pollen allergenicity even further.[63] These conditions alter the proteins in the pollen wall to boost ragweed's allergen content.[64]

And climate change is moving ragweeds to new places. Higher temperatures are pushing its global range from northern France to the Baltic States, and into Russia.[65] Common ragweed is now a major allergen in eighty countries outside of North America, and giant ragweed in forty.[66] Soon it will be impossible to escape ragweed plants or pollen in temperate regions, especially as ecosystems coalesce. Along with storms, floods, droughts, and other disturbances wrought by climate chaos, ragweeds are blooming, evolving, adapting. They thrive on environmental disturbance just as they, themselves, are an environmental disturbance that causes human discomfort on a scale that is impossible to grasp.

In a blink of geologic time, carbon extractions and emissions that underlie global capitalist civilization have changed earth's climate. Increased frequency of floods, storms, and pandemics foreshadow further spillovers, famine, civil strife. Entire species disappear while ragweeds grow stronger, more widely adapted, and more adept at emitting hay fever inducing pollen. Nothing would serve ragweeds better than beefsteak, cars, air conditioning, and endless consumer goods for 9 billion humans. Without question, human progress has made the useless, unattractive, dispersal-challenged ragweeds a remarkable ecological success around the world, a triumph of domination, exploitation, and botanical arrogance: the weed for the Anthropocene. Or Ambrocene.

I sit in front of a computer made mostly in China using rare earth metals mined in places I'll never visit, by people I'll never meet. My computer is powered in part from coal, even though I select "green energy" from the energy provider. Living this comfortable life, I can't escape my contribution to the global proliferation of ragweeds and their pollen. And whatever put these words before your eyes has aided ragweeds as well.

It might seem trivial to mix coarse, raggedy weeds with humanity's greatest global environmental challenge. But both arose from similar notions of progress. Both are motivated by human beliefs and attitudes toward the earth. Weeds are not just plants; climate chaos is not just a weather forecast. They are outcomes of a human-directed global economy that demands constant growth and extraction at the expense of what is left of the natural world. Current winners in this system blindly demand even more of earth's resources. Unstoppable environmental disturbances yield ever more ragweed habitat, seeds, pollen. Opportunity, development, progress—whatever auspicious motives lead to the neglect and denial of nature and of basic human needs have put us on a course leading to further global climate chaos, species extinction, and ragweed weediness. Reflecting on the trajectory and consequences of these interactions, I've struggled to arrive at a conclusion beyond a simple question, something like: Is this what we want?

Giant and yellow foxtail (*Setaria faberi* and *S. pumila*).

8

FOXTAIL

I come to the *Setaria* genus at the end of this book as an offering of a weedy version of hope. The *Setarias* have interacted with people in ways that show a remarkable facility to adapt and change behavior in order to sustain viable populations. They embody a sense of possibility, without which hope is illusory.

Some *Setarias* are ancient grains that continue to provide humans with nourishment. Some cover tropical grazing lands and supply energy for livestock. Others are showy ornamentals. Some species of *Setaria* can be all of these things. A handful of *Setarias* behave both as crops and weeds. And a select few are mostly weeds. They're easy to spot along roadsides and in fields, with their bottle-brush-like tight panicles, some erect, some lightly nodding, and others waving in the wind.

Foxtails, the weedy subset of *Setarias*, followed in the path of crop domestication for as long as humans have fiddled with agriculture. Green, yellow, and knot-root foxtails (*S. viridis, S. pumila,* and *S. parviflora*) became worldwide pests in many agricultural systems. Some biotypes of these same species are still grown as grain crops. Others became weedy in warm-season grain crops like millet and corn or in temperate fields of soybeans, potatoes, vegetables, and clovers.

Giant foxtail (*S. faberi*) is an outlier among the *Setarias*. Tall, robust, and genetically unique, it has all the potential for variable growth but behaves exclusively as a weed, and never as a crop. Not a hint of redeeming value. Unlike other colorfully named foxtails that appear wherever soil is disturbed in agriculture, urban lots, roadsides, and waste areas, giant foxtail intrudes as a major weed in some fields, but not in others. It holds out for the most favorable sites where a particular type of agriculture is practiced. Not every type of agriculture, and not most of the agriculture around the world. Giant foxtail makes its home in the best farmland of North America where intensively managed corn and soybeans fill the countryside. Across the northern plains, from the East Coast to the Rocky Mountains, giant foxtail flashes its shiny green leaves as long bristly seed heads droop in a sort of false humility.[1]

Predictable Assumptions

After too many years of school, travel, and attempts to settle elsewhere, I returned to Ohio in the late 1980s. I came to take a job conducting research on weed ecology. Ohio lands link the Great Lakes to the Mississippi River basin, reaching from the unglaciated Appalachians to the flat Corn Belt's grassland and lakebed soils. This geography would provide the kind of laboratory I wanted for exploring the richness and wonder of life in a species-rich community of weedy plants.

It was a time of tension in Midwestern agriculture, with low commodity prices, farm foreclosures, and industry consolidations. Weeds were part of farmers' discontent—shifts to more troublesome species and regulatory restrictions on herbicides. People were getting tired of fish kills around pesticide-contaminated lakes and streams, and weed-killing chemicals in well water.

Out of thousands of species in the diverse grass family, the one that added most to farmers' uneasiness was giant foxtail. Farmers needed to get it under control so they could produce enough grain to make loan payments, and so they could keep the respect of their neighbors. The herbicide they used was alachlor. It did a good job on foxtails, but alachlor could move into water and was toxic to fish and invertebrates. Farmers saw only two bad choices: lower crop yields and lost income with giant foxtail, or poisoned lakes and streams with alachlor.[2]

As I got to know the local farmers, I learned that many didn't like using chemical weed killers any more than they enjoyed swatting flies. Alachlor, in particular, had an ugly odor and made your skin itch. Farmers sprayed alachlor on the soil before weeds appeared in springtime. It killed germinating foxtails before they emerged from the ground. You never knew when foxtail would start to grow or how many seedlings would be there. So farmers sprayed full rates when the crop was planted, hoping enough alachlor would still be active in the soil when foxtail decided to germinate. It was insurance. Cutting back would be taking a risk because there was no way to predict how foxtail would behave.

Farmers wanted predictions, not a weather forecast but a weed forecast, to tell them when and if weeds needed to be sprayed. Only with a reliable prediction would they risk holding off on herbicide applications. Entomologists had developed prediction methods for certain insect outbreaks, and plant pathologists had done the same for diseases. To do the same for weeds, we would have to assume that the behavior of weeds was predictable. We'd have to forget that these plants were successful weeds because they were not predictable. It was high risk research. But the payoff could be great. Herbicide use could be reduced along with potential environmental harm if farmers would use predictive models to make weed management decisions. And they would save money. That was good enough logic to prepare my research plan.

Just One Word

Giant foxtail is known for its unusual plasticity, the ability of individual plants to change their shape or physiology in response to surroundings. Plasticity is essential for foxtail survival because the generation of novel genetic variation is constrained by a self-pollinating breeding system. It has limited ability to create useful new genetic combinations on which evolution can operate to produce biotypes adapted to specific conditions. Instead, foxtails have a common general-purpose genotype adapted to a range of conditions. Within that genotype, foxtails call on their inherent plasticity to respond to changes in their environment, especially changes imposed by efforts to get rid of them.

Foxtails honed their plasticity over a long period of evolution in variable environments. *Setaria* ancestors emerged from Africa more than 8 million years ago. A few spread to subtropical and temperate areas where they split into about 125 different species. They are typical C4 plants, meaning their photosynthetic system is particularly efficient in hot climates. They take CO_2 from the air and concentrate it around the main photosynthesis enzyme, Rubisco, without wasting energy. *Setarias* can tolerate low fertility and water availability. Early *Setarias* were part of the grassland savanna expansion in eastern Africa that greeted the arrival and wanderings of early humans.[3]

Giant foxtail split off from other *Setarias* in southern China. It came from a rare genetic combination of two diploid (2n) grasses that normally do not cross (probably biotypes of green foxtail and bristlegrass (*S. adhaerans*)). The result was a grass with four copies of DNA, a tetraploid (4n), having four sets of chromosomes, two from each parent species.[4] Over many years of natural selection, a robust line of tetraploid grasses survived as the species we call giant foxtail.[5] It spread across China along with similar-looking green foxtail. Farmers pulled and hoed these grasses from crops of *Setaria* millets, the kind of soil disturbance and seed dispersal that made them successful weeds.

In 1753, Linnaeus had lumped most cluster-flowered grasses into the huge *Panicum* genus. Other taxonomists split off the fine-bearded, foxtail-millet, *Setaria* group around 1812. Other names have been used, but *Setaria* (from *seta* meaning "bristle" or "hair") has stuck since about the 1930s.

The first *Setaria* known to appear in the New World was probably *S. geniculata*. Early wanderers carried it from Eurasia via the Bering land bridge more than ten thousand years ago. This perennial was the first cultivated cereal grain in the Americas, beginning five to eight thousand years ago in southern Mexico.[6] We rarely see it today. It was displaced by maize, a plant of higher productivity, better taste, and greater potential for crop improvement. About forty other *Setaria* species spread to North America, some via South America and a few from the Old World. When Europeans started farming in North America, green, yellow, and knot-root foxtail were recognized as common grass weeds, among the first mentioned by colonists.[7] For almost two centuries, these three foxtails moved with human migration and global trade as typical grass weeds wherever row crops and pastures were planted.

A Place to Roam

The first European to take serious notice of giant foxtail was Ernst Faber. Born in 1839 in northern Germany, Faber followed in the footsteps of his father, a plumber and tinsmith. Somehow, this training led him to enter the seminary at age nineteen. He studied botany and natural history at universities in Basel and Tubingen before joining the Rhenish Missionary Society. Faber departed for China in 1864, four years after Britain and France had forced China to grant access to its interior at the end of the Opium Wars. He joined a cadre of fervent Western missionaries eager for religious persecution. Between episodes of violence and yearned-for suffering, Faber explored the rich flora of central China as the lone German Protestant plant collector with training in botany.

Faber was a prolific writer in three languages, translator of Chinese philosophy, and street preacher. On a botanizing trip to remote mountains in Szechwan Province between 1885 and 1891, Faber collected a specimen of *Seteria* distinguished by fine hairs on the upper leaf surface and a distinctly arched panicle. He packed it among a large collection of specimens that he sent to taxonomists in Europe. German botanist Wolfgang Herrmann sorted through scores of misidentified and improperly labeled *Setaria* specimens. Nevertheless, in the official botanical description, written in 1910, Herrmann named the plant in honor of its collector: *Setaria faberi* (sometimes written *faberii*).[8]

Western countries eventually reaped the fruits and weeds of the Opium Wars. Seeds of what became known as giant foxtail found their way to North America in grain shipments from China during the early 1920s. The earliest recorded specimen was collected September 19, 1925, on Long Island by William C. Ferguson. The tag on the herbarium sheet reads: "Edge of swamp and dump. N.W. of Plattsdale, LI 9.19.25." Ferguson was a Columbia-trained chemist who took an interest in plants late in life and botanized along swampy dumps for personal entertainment.

He was unaware of Herrmann's taxonomic work, so he identified the grass as "*Chaetochloa italica* (L) Scribn.," an old name that led to years of confusion.[9] The next recorded specimens were collected in 1931 by Bayard Long from "disturbed soil and rubbish along the Reading Railway in Philadelphia." The next year, "seeds of a strange millet were found . . . by seed analysts in millet imported from China."[10]

None of them—Faber, Herrmann, Ferguson, Long, or anonymous seed analysts—had a hint that they were looking at a robust polyploid species with the potential to become a major weed across North America. Why didn't anyone raise alarm to quarantine infested sites, halt its spread into the richest farmland on the continent, and save billions of dollars in future control efforts and unmeasurable environmental damage? The answer points to the essential nature of what makes a plant a weed. Giant foxtail, like any species, is not inherently a weed. Blaming its weedy nature and future troublesome impacts on the plant itself overlooks the fervent efforts of its coevolutionary partner and driver of weediness. Giant foxtail became a major weed only because of remarkable chemical innovations, industrial expansion, a cheap food policy, Cold War rivalry, and confidence in unlimited resource availability.

Bayard Long passed his specimens to Merritt L. Fernald, an accomplished botanist, who published the first description of *Setaria faberi* in 1944. He noticed it growing across eastern Pennsylvania, southern New Jersey, Delaware, and into Virginia. It traveled by rail to Missouri and by 1950 had spread from "New York to Nebraska and Arkansas, North Carolina, Kentucky, and Tennessee."[11]

A weed might be a plant out of place, but giant foxtail waited to be a weed until it found the right place at just the right time. The place was the greatest extent of rich agricultural soil in the world, laid bare every year to receive large quantities of weed seeds that would infest fields for decades. The time was the dawn of the age of large-scale chemical agriculture, inaugurated by commercialization of 2,4-D and similar herbicides that kill broadleaf plants (like dandelions, ragweeds and soybeans) without harming grasses (like corn and foxtails).

Giant foxtail had been spotted along field margins in Illinois in 1941 and was soon "locally abundant" in southern Iowa.[12] The timing was perfect: 2,4-D was made available for sale to American farmers in 1945.

Within two years, more than 2.2 million kilograms (5 million pounds) of the stuff was sprayed on farm fields to kill cockleburs, pigweeds, and others without a whiff of damage to foxtails.[13] Giant foxtail's path to weedy success was cleared by farmers spraying 2,4-D. Farmers were encouraged that the battle against broadleaf weeds had been won, unaware that the winner was something they called giant bristlegrass, nodding foxtail, and Chinese foxtail before settling on giant foxtail.[14]

The agriculture industry was slow to recognize this new foxtail species. It resembled larger variants of green foxtail, with a similar though more pronounced nod to the seed head. It was relatively larger, more robust, produced more seeds, and was a better competitor against crops than other foxtails as well as other grasses. Without knowing it, farmers spread the seeds as contaminants in red clover and other crop seeds. Birds, water, and moving soil spread it as well. Under pressure to produce more corn, farmers removed fences and converted the fencerows and field borders—where giant foxtail got its start—into crop production. By 1950, the area sprayed with 2,4-D had tripled, and nodding heads of the new foxtail could be seen from Ohio to Iowa covering soybean fields so densely that crop plants were hardly visible.[15] In a taxonomic study of the *Setaria* genus, Illinois botanist James Rominger described how plowing field borders and extensive spraying of 2,4-D throughout the 1950s had "opened the way for the rapid spread of *S. faberi* [at a rate that] has been little short of phenomenal."[16]

Black Box

As long as I can remember, as a kid pulling weeds from the family garden, I wanted to know how the darn things got there in the first place. Pull or hoe weeds one day, and in a week or so the ground was green with tiny seedlings to replace them. It's a cosmic lesson learned by anyone who gardens or farms. For every weed pulled there are many more just like them waiting as seeds in the soil for years or decades for their turn to take a crack at life in the sunshine. They come from the weed seedbank, the storehouse of seeds on or in the soil. The seedbank is where all annual weeds come from every year. Seeds are essentially tiny, living, respiring plants, an essential stage in the annual plant lifecycle. If the seed rests in the soil for more than a few months, it will spend more of its life in the seedbank than as a growing, green, leafy, flowering, fruiting plant.

When I started to conduct research on weeds in Ohio, I determined to focus on understanding the seedbank. This led me to work with farmers who wanted to get away from repeated broadcast spraying of herbicides. To find better ways to deal with weeds, I needed to be able to predict which weeds would grow in their fields and when they were likely to appear.

First, I needed a way to collect, identify, and count weed seeds in the soil. In all the research papers about weeds, nobody had developed a good standard method to quantify this critical aspect of weed population biology. So I tossed a handful of soil on the lab bench and sat down to pick out the seeds. They were small and dark (when covered with soil), just like a thousand other small dark things in the soil that looked like, but in fact were not, seeds. Gentle pressure caused some to shatter; some were squishy, some firm. What did that mean? I worked at it for an afternoon using a pair of tweezers, magnifying equipment, and lights until I felt dizzy. What I saw most clearly even without lenses and lamps was why most weeds guys had the wisdom to focus their attention on the growing green leafy stage of the plant lifecycle.

Extracting seeds from soil required special skill: tolerance of tedium and monotony. Standard soil sieves were no help except for exceptionally large seeds. Some seeds would float to the top of salt solutions, but so did straw, roots, cigarette butts, and a lot of other things that were not seeds. Eventually I found strong, flexible mesh: ankle-length nylon hosiery in the "women's department" at JCPenney. I ordered a few dozen after a fumbled, contorted, mission-critical explanation of justification to the university accountants. Using the mesh along with a root washing machine gave us a repeatable standard procedure. It required endless hours counting and identifying seeds, but for most species the results seemed acceptable.

The next step was to find out how many seeds of different species were out there in farm fields. That meant sampling soil and extracting seeds from fields of many types around the state. We worked together with research groups around the Midwest to see if we could use the seedbank as a predictor and descriptor of weed populations. Ohio, with its mix of soil types and geography, along with its diverse farming systems, seemed the ideal terrain for exploring how management of fields and farms influenced the seedbank.

On bright spring mornings in the early 1990s, my research crew—a couple of technicians, graduate students, and undergrads—would pack up a truck with buckets and bags and tags and maps and soil samplers. With coffee, donuts, and promised lunch at La Fondita, I convinced them to drive into the crisp air to take soil samples. When we showed up on farms with a university truck, all we had to do was mention weeds and we got a free pass to fields, pastures, barnyards, or wherever we wanted to go. Every weed seed in the state seemed to belong to us, a boast that went unchallenged.

Back at the lab we unloaded hundreds of bags of soil. The relative fun of digging soil samples in the fresh air was followed by hours of screening and sifting and subsampling and counting. Every seed and seedling had to be identified to species. Months later, data from napkins, scraps of paper, and computer files were assembled, and we tried to make sense of them.

We looked first at data from large-scale conventional farms, the most common on the landscape. Those with a long history of corn and soybean production had unusually high numbers of giant foxtail seeds, sometimes more than 49 million seeds per hectare (about 20 million per acre). Farmers could never kill all those foxtail plants; a few would always survive. And each plant can disperse more than ten thousand seeds. Even a few stray foxtails could build up a large seedbank in a few years. And those seeds persist, some up to forty years.[17] It didn't seem to make sense: the farms that used the most mechanization, fuel, and herbicides were the ones with the most giant foxtail. Farmers who put fields into alfalfa or wheat, even for a single year, had fewer foxtail seeds. But conventional farmers didn't make money on alfalfa or wheat; only high corn and soybean yields would make their loan payments.

Knowing how many weed seeds were in the soil was only the first step to understanding dynamic changes in the seedbank. We also needed to untangle the connection between dormant seeds, nondormant seeds, and seed germination. Dormant seeds are healthy and viable, yet they refuse to germinate even under ideal conditions. Dormancy protects a plant's precious reproductive resources—the seeds—against harsh conditions in the soil where they are exposed to a hodgepodge of temperatures and water levels. If they didn't have dormancy, seeds might germinate at the wrong time and emerge when conditions for growth and reproduction were unsuitable. Dormancy gives weed seeds time to move with animals, water, and wind, or to rest in the soil for many years until the time is right to germinate. Crop seeds don't need dormancy because humans collect and store them in conditions that ensure uniform germination after planting the next year.

If all the seeds on a foxtail plant had the same degree of dormancy they would wake up and germinate at the same time. That would make prediction easy, but it would not be a good strategy for a weed to leave all its progeny vulnerable to control at once. Foxtail devised ways to allow some seeds to germinate soon, and for other seeds to wait many years in soil. This spreads germination opportunities out over time. Foxtails do this with a type of physiological plasticity, producing seeds with several conditions of dormancy. Some foxtail seeds are deeply dormant, some conditionally dormant, and some are not dormant at all. Deeply dormant seeds, like deep sleepers, require unusual conditions, like a long cold period, to awaken from dormancy. When seeds begin their release from dormancy, they become conditionally dormant and germinate only under a narrow range of conditions. Some of the seeds that are initially nondormant can switch back into a state of secondary dormancy.

With all these different levels of dormancy, random groups of foxtail seeds can germinate, and seedlings emerge, throughout the growing season. If early emerging seedlings are destroyed by tillage or weed killers, there's another cohort, or two, or three, ready to replace them. Some seeds become nondormant, poised to germinate, but high summer temperatures send them into secondary dormancy. They can go in and out of dormancy with annual temperature cycles.

When a giant foxtail seed ripens on the plant, it can be in any of these dormancy states. Two seeds right next to each other on a bristly panicle are likely to differ in their ability to germinate when they fall off the plant. This unusual germination plasticity is the key to ecological success of giant foxtail. By producing seeds with variable dormancy, foxtail places bets on different outcomes to guarantee a favorable result at the population level. It's a hedge-betting strategy that reduces the risk of population decline by supplying the seedbank with enough seeds that are prepared for any environmental turn of fortune.[18] A large group of seeds germinates early, giving plants the whole season to grow and reproduce. Other groups of seeds germinate throughout the growing season whenever moisture and other conditions are good for growth. Or they can wait it out until some later time. The result is an annual rain of foxtail seeds with a wide range of germination and dormancy behaviors. Another result is that foxtail seed biology is very difficult to study, and even more difficult to predict.

Foxtail Forecast

With all their flexibility in dormancy, over half of the foxtail seeds in the upper soil germinate in a surge of optimism at their peak time in spring— a time known only to them. The freshly launched foxtail germlings are skinny and fragile before their pointy little leaves poke out of the soil. Energy stored in the seed is depleted to push the tiny white and wiry seedling into the sunlight. Until then, these tender "white stage" germinants are easy to kill with simple tillage or minimal rates of herbicide. If we had a way to predict when that susceptible stage was about to occur, farmers could optimize the timing of their control measures, using shallow tillage or lower doses of herbicides. It would save money, time, and unnecessary herbicide treatments.

To predict the peak germination time, we needed to understand what set these seeds off on their weedy destiny. What was the specific environmental stimulus for germination? Many weed researchers around the world were trying to figure this out for one weed or another. The answer would be a step toward developing mechanistic models to simulate and predict the behavior of weeds. The digital universe was looming, and suddenly applied biology wasn't science unless it was coded in C++. I left the coding and modeling to others and focused on seed germination and emergence. The research assistants and students did the tedious stuff, conducting dozens of experiments on seed germination responses to physical and chemical variables.

Foxtail germination time, pace, and consistency were controlled by environmental conditions. This was the case for most other species. But for foxtail, the germination behavior changed from one population to another. Sometimes from one plant to another. The changes began in the mother plant and continued throughout seed dispersal and storage. Finding a reliable signal for germination seemed impossible. Giant foxtail seeds didn't respond to light, dark, or physical manipulation. They didn't follow a regular pattern with temperature: the same temperature that stimulated some foxtail seeds to germinate, suppressed others. The seeds shifted from one state of dormancy to another, sometimes during the middle of an experiment. If the soil was dry, high temperatures released seeds from dormancy; if soil was moist, high temperatures put nondormant seeds into secondary dormancy. Moisture stress, or alternating wet and dry conditions, stimulated germination of some seeds but put others into a state of dormancy. After a few lab experiments with foxtail seeds I couldn't see how its shape-shifting behavior would possibly be reduced to a predictive model.

Luckily, other research groups were able to make some sense of foxtail seed dormancy and germination. Ingenious studies by colleagues in Iowa and elsewhere showed that giant foxtail seeds germinated in response to a combination of water, temperature, and the amount of dissolved oxygen in the water they imbibe from the soil. Water and temperature made sense. But regulating the amount of O_2 that made its way inside the seed required a springtime "air pump." The amount of O_2 dissolved in water changes with temperature. Colder water holds more O_2. As cool, moist soil is warmed by the rising sun, O_2 becomes less soluble and escapes into the air. In this way, cells inside a foxtail seed get a daily pulse of O_2, which is trapped inside by cell membranes and walls. This springtime O_2 oscillation is the signal that tells nondormant foxtail seeds that conditions are right for germination.[19] It is a marvelous thermo-hydro-oxy-driven process, the kind of thing that passes for poetry among weed seed researchers.

By the year 2000, computer models had been developed that used field temperature and rainfall data to predict germination and emergence for a few weed species. I assumed, of course, that farmers would find this useful. So I showed how it worked at a local extension meeting. In spite of their respectful lack of enthusiasm, the results for broadleaf weeds like pigweed and velvetleaf convinced a few farmers to offer to collect temperature and rainfall data and let us count weeds in their fields. Then Dwayne, a farmer I had worked with for years, raised a hand. "Can you do this for foxtail?" All I could say was that good predictions for the initial burst of giant foxtail germination would require more complicated data. A model based on air temperature alone gave a prediction as good as a wild guess. Adding soil temperature gave a better estimate, but that required a special thermometer and daily monitoring (with the technology available at that time). There was no easy way for him to measure available water. Some mysterious oscillating oxygen pump? Hardly worth mentioning.

"You might predict emergence for one or two weeds," he offered. "But you know my farm. There's a dozen different species in my fields. You'd need a prediction model for every one of them." I stumbled around for an answer. It occurred to me that we already had data on twenty or so species, and there were patterns among them. Giant ragweeds were reliably early, and morning-glories reliably late. And the sequence of germination of all the weeds in between was about the same every year. Maybe by predicting one correctly, I suggested, we could predict others. Dwayne shook his head: "Maybe. But not giant foxtail."

As One Rambles

I spent a lot of time driving the countryside monitoring fields and weeds. I made follow-up visits to fields where we had taken samples of the weed seedbank. I recorded data to see how well the weeds in fields corresponded to what we had found in the seedbank. I also made note of the greening of pastures, the first coltsfoot blooming yellow along the roadcuts, and when a particular weed first appeared, flowered, or set seeds. I recorded differences between the north- or south-facing slope, the wet spots and the ridges, fall-plowed and spring-plowed fields. My small pickup became my office, a place to study patterns that might be instructive.

To accompany me on my drives I found a collection of recorded books at the public library. I pushed a squeaky cassette tape into the player and listened to *Walden*, Thoreau's musings on nature. The narrator, William Peirce Randel, will, in my head, forever be the even, plodding, unemotional voice of Thoreau. I drove the rolling hills while he explicated, unhurriedly, the rivulets of water from puddles after a rainstorm as fractals of the arrangement of streams and rivers at larger scales. It was like hearing the Farmer's Almanac in nineteenth century prose. I expected that Thoreau might spout aphorisms about the best time to dig a post hole, put up jam, or build a fence. Instead I was roused by verbal phrases unforgettable to those of my persuasion: "Shall I not rejoice also at the abundance of weeds whose seeds are the granary of the birds?"[20]

In his journals, Thoreau recorded the flowering times of 465 plant species in the area of Concord, New Hampshire, from 1851 to 1858. He kept track of these things, he said, in an effort to live according to the cycle of nature's schedule.[21] First flower or full bloom of a plant were indicators along the progression of the biological calendar that moved with the turning of the earth and its path around the sun. Thoreau was describing phenology (fe-NAH-luh-jee), the timing of events in plant and animal life cycles in relation to changes in the season. The times of bird migration, egg hatching, wildflower blooming, insect molting, and other phenological events have been observed and recorded for centuries.

Thoreau did not take note of emergence or flowering of grasses, and only mentioned a few weeds. But farmers had connected weeds to phenological events for a long time. Maybe I could link the patterns of the natural world that Thoreau was describing with phenological events associated with weeds. Like germination and emergence of foxtail. I already had a notebook full of records for the first flowering of wild plants and of several weeds. Like all plants, weeds integrated the effects of temperature, water, and air in similar ways to initiate and terminate growth. They might follow similar patterns of change in the seasonal progression from early spring into summer when most annual weeds germinated and emerged. And Dwayne was right: the most challenging weed to test this idea would be giant foxtail.

So I went to see Dan Herms, an entomologist and expert on native and ornamental plants and their insect pests and pollinators. He studied phenology to predict insect pest activity. It was second nature to him that bronze birch borer adults start to emerge just as black locust begins to bloom; egg hatch of pine needle scale coincides with full bloom of common lilac. He was a phenology encyclopedia in street shoes. He had followed research that updated data collected by Thoreau, showing that due to climate change, plants in Concord now flower a week earlier than they did in the 1850s. Best of all, he was willing to share data on flowering phenology to see if we could predict foxtail emergence time.

Giant foxtail was abundant across the landscape and its germination and emergence patterns were too complicated to predict using rainfall and temperature data that were easy to collect. Would it be possible to fit a variable event like giant foxtail emergence into the natural sequence of flowering? Dan was already collecting the data on first and full bloom of woody plants to predict insect development. It was not a simple task, nor was it unpleasant, walking the woods every day from late winter to mid-June for four years to observe buds and count flowers. Distinctly less pleasurable was the task of keeping track of the foxtail seedlings emerging from the soil. The first leaf of an emerging grass seedling that pierced the soil surface was indistinctly tapered, pointed, and green. In other words, they all looked alike. Only a trained eye could distinguish giant foxtail from other weedy grasses.

Data we collected over a couple of years showed that foxtail seedling emergence fit into the biological calendar surprisingly well. And the results were practical: when the first red chokeberry flower opened, foxtail emergence began. When multiflora rose reached full bloom, 80 percent of foxtails were out of the ground. It didn't matter if it was an early warm spring or a late cold one; the sequence of flowering and weed emergence remained about the same. This regular order of flowering events provided a progression that farmers could follow to anticipate foxtail emergence times. Overall, predictions using flowering plants were on target more than predictions based on temperature and moisture. And it was simple. Without complicated measurements and computer code, we needed only to look at the plants and flowers of fields and forests.[22]

It seemed like we finally had a simple approach to prediction. I reported our results in 2007. Hardly anybody paid attention. Perhaps this, too, was predictable. The worldview had shifted. The value and reverence for observing the natural world had declined, replaced by looking at a screen. The culture of chemical farming and GMO crops had reinforced more herbicide dependence, not less. The interests of farmers, crop consultants, industry, and university researchers had turned from prediction models to more high-tech issues like herbicide-resistant weeds, an elite faction eventually joined by foxtails.[23]

To Find the Root

On one of my weed monitoring drives, I took a turn down Apple Creek Road from Wayne into Holmes County, Ohio. It was early July, shiny yellow buttercups bloomed, and a fresh cutting of alfalfa dried on the hillsides. Slowly, the scenery shifted from mostly conventional corn-soy "English" farms to Amish horse agriculture. I finished my field notes and walked across a pasture to a high spot where I could look across one ridge after another toward the blue haze of the horizon. I leaned against a fence post and took note of differences in the crops, their uniformity, and field size that distinguished Amish and English farms. I tried to read the landscape, what it might tell me about the view of the world through the eyes of those whose minds were reflected in the shape, size, and arrangement of the patchwork of fields, pastures, and woodlots. On a typical Amish farm, shocks of barley stood in rows beyond the fence line. Fields were small, with wide borders, a mix of hay fields, vegetables, cow pastures, and somewhat ragged field corn. An "English" farm stood out across the way with a large soybean field marked by lines where a herbicide sprayer had recently passed.

More significant were the weeds. Foxtails were bio-socio-indicators, sensitive enough to their surroundings that different species pointed to different cultures. Specific foxtail species represented different environmental conditions imposed by different historical and current farm management practices. At root, they signaled different assumptions among managers of these fields based on corresponding beliefs about the environment and the natural world.

Giant foxtail broke through the crop canopy on English farms where a tractor-powered high-density precision planter left its track. This species seemed to thrive in soil compacted by heavy equipment on fields worked early and wet in springtime. In those conditions, giant foxtail would beat out any other grass. Giant foxtail dominated fields in agricultural systems that featured continuous GMO corn and soybean production, herbicides, inorganic fertilizers, and fossil fuel power. Likely, these farmers used the same mix of herbicides year after year; it did a great job on most of the weeds, save a few to pass along genes for increased tolerance, and left the crop looking uniform from one end of the farm to the other. I knew from samples we'd taken here that the seedbank in these fields had seeds of over thirty different weed species. But among the emerged species, giant foxtail thrived above the others, the biological marker of resource extraction on behalf of continuous growth.

Amish horse agriculture had low corn populations, slightly uneven rows, and cultivated soil between the rows. Four-year crop rotations with clovers and grains and somewhat later planting dates left the soil in a condition that favored yellow foxtail, which grows better in lighter soils and germinates later than giant foxtail. Amish farmers in this area use few to no herbicides. Instead, they rely on ingeniously designed horse-drawn mechanical cultivators. The result was a mix of several species. No single weed dominated or gained tolerance, and, of course, there was no selection pressure for evolution of herbicide resistance.

There's no moral or ethical difference among foxtails or other mixes of weeds that I'm aware of. Foxtails were simply signals of biological selection. One approach to farming left fields infested with a solitary robust species selected by and demanding increased herbicide spraying. Another approach to farming favored greater biological diversity of less competitive and more easily tolerated species.

Other differences appeared on the landscape. Most of the English farm fields here were owned by one of two farmers who didn't live nearby. They contracted with the local grain elevator to supply, plant, spray, harvest, dry, store, and sell the corn and soybeans. They weren't farmers so much as miners, extracting nutrients from the earth and selling them to distant buyers. The Amish farms, by contrast, were tended by individual farm families who lived, shopped, auctioned, banked, and worshiped in the community to which they and the land belonged.

In spite of their plain dress and horse-drawn technology, the Amish hosted a mix of modern innovation and entrepreneurship in organic, biodynamic, and other alternative management approaches. Their agriculture evoked a Jeffersonian ideal of a nation of farmers. The scale was generally small—the extent that one person could manage—and community interactions and support were essential. Here stood humble possibility, with remarkable enterprise and prosperity that sustained and supported an entire community, including the land, wildlife, and people—with their flaws and inconsistencies. The scene fit the model of the American free market system as well as any agricultural enterprise. Except for one thing: they recognized limits—of resources, needs, and growth on a finite planet. Within those limits were agricultural practices that emphasize renewal, diversification, and internal cycling of resources.[24] Weedy foxtails were there, too. Not a lone imposing species, but a diverse mix along with other weeds that were accepted as mere inconveniences.

Where Meaning Lies

Our project on the seedbank and giant foxtail emergence wound down around 2012, and I prepared to submit the final report. Maybe the whole idea had been crazy to begin with, that farmers might put their trust in chokeberries or multiflora roses to aid in management decisions and risk whole fields covered with giant foxtail. The regular sequence of flowering of woody plants and timing of weed seed germination were unlikely to change farmer behavior and their dependence on routine herbicide spraying.

The disconnect between the novel use of plant phenology and the reality of agricultural practices faced me when I sat down to write the required impact statement for the project. I could report that if farmers all across the Midwest adopted our approach, they could—theoretically—save millions of dollars by reducing the amount of herbicide sprayed. I estimated that if farmers using the biological calendar eliminated even just 1 percent of herbicides, they would save $50 million per year. That would translate into unaccountable savings in downstream health and environmental benefits.

A different perspective on the impact of this research would be that for a brief period, a few researchers entered into nature's time scale. We let flowering plants point the direction and set the agenda. We stood by our understanding of temperature-dependent enzymes that control biological activity. But we dipped a foot into other ways of knowing, other ways of seeing the world. We spent mornings on our knees, fervently counting foxtail seedlings that emerged during the night. We walked through the woods to record the unfolding of buds and flowers. We couldn't help noticing the changing color of moss on the trees, the first appearance of a leaf miner, spring peepers chirping in the swamp, the changing chorus of birds watching from treetops.

Whatever mechanism moved this along—even an oscillating O_2 pump—was inconsequential in the changing display. The stretching of tiny foxtail leaves to reach above the soil surface and resume photosynthetic activity in the life cycle of the weedy grass was connected to the flow of sugars through trunks of red maples to the tallest leaf buds whose disclosure signaled the regular appearance of buds, flowers, leaves, insects, birds, spiders, frogs, and others joining in the unfolding of springtime.

With the project nearly over, I took another drive down Route 83 to look at fields and weeds and flowering trees. Thoreau droned in the background. Winding through the hill country past field after field of tasseling corn, I sensed an uneasiness in the richness of the crop and the general poverty of the surrounding population. In all my study of agronomy at the great Land Grant universities I attended, the overwhelming assumption was that "agriculture" was the conventional and mechanized production of crop commodities. We assumed this system was "needed" to "feed the world." We assumed that a crop "farm" was a progressively larger and increasingly specialized operation. We assumed this is how farm families would prosper. The fields outside my window produced mostly corn and soybeans in a chemicalized, genetically manipulated, and digitized system that fed confined animals and made high-fructose syrup and ethanol based on practices prescribed by the boards of directors of industry conglomerates. And the old assumptions had gone unchallenged and persisted on this land.

I turned off the main road, away from Kokosing and out of the Muskingum watershed. In my travels to collect seeds buried in the earth I had crossed through places with names like Powhatan, Tontogany, and Piqua, from the Wahkeena Reserve to the Tecumseh Natural Area, and along rivers called Mami, Scioto, Maumee, and Olentangy. The names were reminders that this land was once home to other cultures with different perspectives of their relationship to the earth, to nature, and to each other. These were cultures whose respectful ways inspired people like Benjamin Franklin and other founders of American democracy.[25] For centuries, they lived in these same valleys and hills without destroying mountain tops, ravaging wildlife habitat, polluting shared resources of water and air. They farmed these soils and sustained these lands for generations with different perspectives on wealth, resource extraction, social equality, food, and agriculture. And weeds.

South of Millersburg, the Walhonding and Tuscararwas rivers opened to a wide valley before joining to form the Muskingum River that flowed to the Ohio. This was one of the most fertile valleys in the unglaciated part of the state, where alluvial soils formed from river outwashes. Somewhere in my notebooks were data from seedbank samples taken along this watershed.

This region intrigued me because my family had recently sold our run-down farm in the hills about twenty-five miles east. I had walked this landscape over the years. The whole river basin had been settled by the Lenape (Delaware) people after they were run out of eastern Pennsylvania by European settlers. The valley was surrounded by upland maple, oak, and elm forests, resources and an ecology that were familiar to the Lenape. Reaching out from their highly organized settlement at Gekelemukpechink (Coshocton), they developed a sophisticated economy and civil society. They cleared the land slowly, using rudimentary hand tools. In the openings they planted maize in May, according to their biological calendar, when the hazel trees bloomed. Maize and other crops expanded to thousands of acres across this bottomland without mechanization or even plows. They produced enough for themselves and for trade across several watersheds. Mounds of hand-planted open pollinated maize, climbing beans, and creeping pumpkins, squash, and melons—complementary in the field and on the table—flourished according to an ancient system.

Weeds must have been a feature of that system. The best information I've been able to find suggests that the mounds were kept in place and weeded (by women) after planting. The maize emerged first followed by beans and then ground-covering pumpkins and squash that filled in around the hills. After a few years, production moved to other fields, possibly due to depletion of soil nutrients. And due to weeds. According to David Zeisberger, a Moravian missionary historian who lived and worked among the Lenape between 1767 and 1780, "When . . . their field begin to grow grass . . . they leave them and break new land, for they regard it as too troublesome to root out the grass."[26]

So weeds—grasses in particular—disrupted the three sisters. Not the foxtails we see on the farmscape today, but species like *Panicum, Danthonia, Andropogon,* and others.[27] Green, yellow, and (probably) knot-root foxtails didn't appear in this valley until Europeans again displaced the Lenape. The deliberate approach to crop production among the Lenape was in keeping with their worldview based on a notion of reciprocity. They would give back to the land for what was taken, to replenish the earth that sustained human life. In their resilience they built new homelands and developed bonds with the land wherever they were forced to move, by foreign invaders or by troublesome grass weeds.

I stopped at the fields where we had sampled the seedbank and noted the weed species that showed themselves between the corn rows. I tried to picture the aggregate of the living species before me—those showing themselves in midsummer along with dozens more weedy species in the seedbank, all connected to hundreds of associated insects, fungi, bacteria, nematodes, and other organisms. This was the art of looking at weeds, linking the plant community to the greater biological community and ultimately the human community.

Some of the weeds in these fields were distant progeny of plants that had been hoed or abandoned by people who occupied this land many generations before. My seedbank samples represented the history of weed genotypes that had cross pollinated over many years. Their genetic history reached back and expressed something of the minds of people who've tended this land. Few of the species in the seedbank of these fields— lambsquarters, amaranths, witchgrass among them—had genetic histories shaped by the native inhabitants of this land. Most of weeds now in the fields were introduced from elsewhere, with genetic histories going back to Europe, Asia, and a few to the far West and South America. Modern local practices imposed agrestal selection with fossil fuels, mechanization, and pesticides that left lower overall biological diversity than would have been found among wild populations of ancient farming systems. As a result, the favored species were novel foxtail-like genotypes and a few broadleaf companions.

Hope Reconsidered

I began this chapter with the suggestion that *Setarias* might tell us something about hope reflected in botanical plasticity and human possibility. I'm not unrealistic about the social and environmental despair I see in the Ohio countryside where I grew up and eventually returned to study foxtail and other weeds. My own first step in response to all that has been lost of the natural world and of community is to find reasons to believe that alternative ways of being are possible, based on awareness, understanding, and acceptance.

Thoreau put his faith in a seed, an apt symbol of hope and potential for transformation. He questioned the basic assumptions of his time and expressed a belief that he, like every individual, was obliged to determine right from wrong in order to live a spiritually meaningful life. While some of his writings are distinctly curmudgeonly, Thoreau's musings on seeds celebrate abundance, fruitfulness, and human connections that underlie communities and give hope to new generations.[28]

Foxtails put their hopes for weedy success in the physiological plasticity of their seeds. These remarkable units of reproduction give them resilience, an ability to make it through tough times, to avoid the too hot, too cold, too wet, too dry. To live when and where conditions for life are good. And in these seeds, of course, are the genetic instructions for doing so. But even the substantial plasticity of foxtails has limits. They are plants after all. They can only undergo significant adaptive change through evolution involving gene recombination and selection. This takes time, sometimes many generations, even for a severe selection pressure like repeated herbicide spraying.

Foxtail's coevolutionary partners, by contrast, have enormous potential plasticity. Their reasons for hope should be commensurate. In particular, hope springs from the innovation and vitality of hill-country communities that thrive outside the broad patches of what Mary Oliver called the "frightened, money-loving world." These communities operate within the biophysical means of nature, where economies are based on trust, conviviality, and tolerance of a few inconvenient weeds.

The natural world itself is reason for hope. The patchworks of green that persist along waterways, fens, and hillsides still out of reach of industrial farming and mining are nature's flag of resistance to human manipulation. The tenacity of flora and fauna adapting to environmental change demonstrate the resilience of the natural world in spite of increasing habitat fragmentation and other pressures. Flows of energy, materials, organisms, genes, and ideas within and across landscapes operate at scales of complexity beyond any tractable models.

Giant foxtail as a dominant grass weed throughout this environment was not preordained. Where attitudes about land and resource use differ, it is still not a major weed. This is why *Setaria* signifies the possibility and hopefulness of choosing different paths. Hope seekers don't have to accept that what dominates the social, cultural, and farm ecology today has any meaning other than to serve as a caution. There are other possible outcomes. They depend, as Thoreau suggested, on what individuals do with those possibilities: how we farm, how we eat, how we consume, how we behave with each other and with the natural world. These choices will determine who we are and what our weeds will be.

EPILOGUE

What's 'Round the Bend

Wherever people have lived, played, and moved, weedy plants of one species or another have poked their way in to spoil the fun and fruition. Their natural histories and ours are inseparable. Some are ancient annoyances, others a recent nuisance. The ones that became highly successful botanical troublemakers did so by entangling human creativity, imagination, and contradictions against a backdrop of the natural world.

The ancient struggle with weeds connects us today to other struggles—environmental, social, ethical. Wayward plants are at the heart of public, and politicized, concerns about the food system. Weeds cost society billions of dollars in lost farm productivity, higher food prices, hazards to health and safety of people and animals, on top of expensive control efforts. Up to now, the blame for all this has focused on lowly botanical creatures that operate by nature's basic rules of evolutionary biology. Clearly, hapless herbs couldn't evolve the capacity to instigate such social, environmental, and economic disruption. Only the assistance of coevolutionary partners could entangle mere plants in such turmoil.

To better understand the nature of our weedy dilemma, I've tried to explore the underlying agency, the mingling of human attitudes and behaviors with plant behavior and underlying biology. I found that faultless flora isn't made into vegetal vermin by devious means. Most often, people have been motivated by good intentions—to be a good neighbor, pay debts, forestall another Dust Bowl, and solve the problems created by yesterday's solutions. To feed the world. To keep the shareholders happy.

We live with weeds in a world of ambiguity where everything has been touched by human influence. What used to be thought of as "nature" has been irreversibly altered by carbon overload, biological invasions, and species loss driven by habitat destruction from agricultural, industrial, pharmacological, military, and other activities. A gene-edited world is being thrust on us. Our biological and technological future has never been more uncertain. In researching these natural histories it's been instructive, if not especially comforting, to look at the history of social and environmental consequences of well-intended efforts to employ technology to defeat weedy aspirations. How did we get to this place where the next turn of the screw may be an existential threat to earthly habitation? I researched and wrote these natural histories thinking some insight might be found in the outcome of ordinary plants interacting with ordinary humans in their ordinary ways.

A Plague in the Year of the Rat

This book was conceived and (mostly) written before SARS-CoV-2, the causal agent of coronavirus disease (COVID-19), reshaped the way we think about and act in relation to each other, the global food system, and the natural world. To the surprise of many, a strange and previously unknown RNA virus transformed human activity on most of the planet. Yet our ongoing entanglements with weeds might have been a reminder that novel coronaviruses and invasive species generally have origins in human misunderstanding of science and mismanagement of the natural world. Thousands of wild plants are around us and most are harmless; many are essential. Weediness occurs when humans disturb their environments, move them around, reduce their competitors, change resource availability, and bring them into closer contact. Likewise, millions of viruses are around us and most are harmless, a few essential. Spillover from animals occurs when humans disturb alternate hosts, kill off predators, alter habitats, and inadvertently bring them into closer contact.

Ignoring basic evolutionary biology will keep us in a state of threat from the next spillover or pest outbreak event. We've seen this in the natural histories of a handful of plants that followed the paths to weediness and exposed our entanglement with the results of our own behaviors. What is strange is not the organisms or their biology, but the human response, beginning with denial, blame, and ultimately neglect of fundamental social and economic changes that might keep it from happening again. Or that might prepare for what comes next. The pattern is not unfamiliar in farming communities facing major outbreaks of plant pests that continue to evolve pesticide resistance.

No weed is likely to provoke the level of suffering that a viral disease can cause. But over the history of agriculture, weeds and their control have stolen the time, energy, productivity, and well-being of millions of people. These stories illustrate that weeds are still doing so. The social, environmental, and health impacts of extensive herbicide spraying, in particular, have yet to be tallied. My point is not to compare miseries, but to encourage greater appreciation of how weeds, pest insects, plant diseases, and virus pandemics happen at the intersection of evolutionary biology and human behavior.

The pandemic experience also tells us something about resilience—ours and that of the planet. A few months into the global health crisis, clearer skies and cleaner waters inspired a sense of possibility. Air pollutants, fuel consumption, and traffic noise declined. The earth took a breather, for a brief while, until the push for "recovery" meant resuming the same hectic activity and resource exploitation that had disrupted natural systems and made a virus outbreak inevitable.

Living in a pandemic taught us that we have incredible plasticity. Of course, adaptability alone is not necessarily reassuring. Millions of people survive with remarkable adaptability in wretched poverty far from anything resembling nature. But as our inherent plasticity allowed us to behave differently in a pandemic, we might use that same plasticity to treat the earth and each other differently in whatever follows. Not just because of global threats from disease and climate chaos, but because we know there are better ways for all people to live.

In Defense of Nuance

Critiques of big agriculture are easy to find. Alternative approaches to agriculture have their share of detractors as well. The topic of GMO technology, in particular, seems to be cast only in extremes—sustainable or unsustainable, safe or dangerous. My discomfort with the current state of agriculture is clear, along with my entanglement in the techno-industrial agricultural system. I hope this exploration of the nexus of biology and history encourages readers to consider their own connections to agriculture, weeds, and the natural world.

Beyond emotional issues are larger questions of how to sustain food production in ways that respect the natural environment while providing more healthy, fair, and mindful ways of living for increasing populations of urbanized people. If field crops remain central to food production, weedy plants will be there, too. Nobody wants to spend their life bending over pulling weeds, a timeless dilemma still in need of solutions with the fewest negative impacts on the environment and society.

Industrial agriculture will change when consumers make change happen. What consumers are willing to pay for, farmers will find ways to produce. We vote about the sustainability of food production every time we sit down for a meal. One could vote at the kitchen table against environmental and ethical problems associated with confined animal feeding operations. The choice of what to eat or not eat will influence the industries that dictate federal food policy. Scientists can vote about the kind of research they conduct. A vote against research on herbicides and corn production can open other areas of study that might be more interesting and ethically sound. Everyone must make their own choices. There's abundant information available to guide us from a scientific perspective. The only feasible approach for how to organize and feed ourselves in a more equitable and environmentally responsible way can be found in the mirror.

Hope and Good Intentions

Where there are people there are weeds. This is the great truth we can't escape. Any hopes for a future where extractions, emissions, populations, and food production better balance the earth's biophysical capacity to sustain a fair and resilient society are also hopes for future selection and adaptation of weedy plants. Creating truly democratic and just economic systems that respect and revitalize natural systems will change agriculture in big ways. There will be impacts on crops, soils, and, inevitably, the agrestal selection pressures to which weeds respond with remarkable plasticity and capacity for genetic change. New paths to weediness will emerge as plants respond to changes in resources and environments. Humans, in turn, will respond to their evolved changes, and this will exert other, new, selection pressures to which weeds will continue to evolve. A current buzzword among agriculturists with an eye on future production is "climate-smart" agriculture featuring drought- and high-temperature-adapted crops. Of course, the evolution of climate-smart weeds is already underway.

It might seem unlikely that modern society would turn the Anthropocenic ship in another direction. But the initial response to the global pandemic caused by COVID-19 proved that the way things are—or were—can change. The histories of weeds tell us this as well.

However we decide to live, farm, and eat in the postpandemic world, we can be certain that some of today's troublesome weeds will become insignificant as future agriculture rolls along. Likewise, leafy waifs that we now consider unimportant will follow us along new paths to weediness and emerge as tomorrow's major pests. The ones to look out for are species that have been transported, sold, shared, fertilized, watered, coddled, and nurtured by human interactions. In other words, plants that attract our attention and call to us in practical or emotional ways. Plants—desirable or despicable—will change due to evolution through genetic and epigenetic changes in response to human-mediated alteration of their environment. And global movement of plants will continue to shuffle the biological cards and potentially bring some truly noxious species—loved and admired in some other context—to new crop production regions. These species will survive because of their plasticity and ability to evolve in new conditions. Resilience of plants reflects the resilience of nature, in which we, and weeds, are inextricably entangled.

These patterns persist because weeds remain botanical expressions of human nature, the outcome of ancient and ongoing interactions among plant life and people. With new ways of crop production will come new ways of weediness. Genetically based plant traits, including dormancy, dispersal mechanisms, and resistance, will work along with plasticity to make malleable species. Their success as weeds will rest with their coevolutionary partner and characteristic human traits, including greed, shortsightedness, laziness, gullibility, fascination with technology, hubris, and others. Where there are weeds there are people.

Of the thousands of species in the Plant Kingdom, weeds represent a small subset, one with which humans interact intimately. We're engaged in an ongoing coevolutionary dance with weeds, a tango of frustration and reaction moving down paths to even greater weediness. We've tangled for millennia with this tiny fraction of the living organisms on the planet. The result has not always been favorable for us or the planet. The evidence is all around us in fields and farms and cracks in the sidewalk. So maybe a little humility is in order. We've been outsmarted by weeds, after all. Maybe it's time to be a little less pretentious in our biological tinkering, a little more modest in expectations of what our brilliant technology can do. Maybe we can show a little more appreciation for what nature has done and can do, and a little more respect toward the adaptability of particular green leafy life forms that continue to evolve in ways known and yet unknown.

NOTES

Introduction

1. I've used common names accepted by the Weed Science Society of America (WSSA), some of which differ from those in the USDA PLANTS database. The Integrated Taxonomic Information System resolved conflicts among scientific names. Historical names are used when discussing a time before modern names were invented. For example, I used "Taraxacum" and "Abutilon" to describe these plants before they were known as dandelion and velvetleaf. Confusion of weed names and identity is part of their stories.

2. Emerson, Ralph Waldo 1878, 3.

3. Blatchley 1912, 6.

4. Baker 1965.

5. These stories are about weeds, not invasive plants. Weeds are maintained by human disturbance (digging, plowing, tilling, mowing, spraying, etc.) in flowerpots, home gardens, lawns, waste areas, and large-scale agriculture. Invasive plants persist in "natural" areas (e.g. forests, woodlands, swamps etc.) with little human intervention. Plants can be both weedy and invasive, depending on how people disrupt their environment. My distinction is open to dispute; I will muddle it for ragweeds in highly modified industrial ecosystems.

6. Harari 2015, 94.

7. Human enslavement has long been connected to agriculture, but its origins also included mining, monument building, and other less-valued menial labor.

8. De Wet and Harlan 1975, 99.

9. Darlington 1847.
10. Selby 1897, 275.
11. Van Wychen 2018.

1. Dandelion

1. The sequence follows Richards (1973), but the date follows more recent analysis by Panero et al. 2014.
2. Panero et al. 2014, 49.
3. O'Connor 2008, 147.
4. Van Dijk 2009, 48; Wright, Kalisz, and Slotte 2013, 7; Noyes 2007, 207.
5. Richards 1973, 190.
6. Mitich 1989, 537.
7. Several species of *Taraxacum* are called "dandelion." Not all those mentioned here pertain to the apomictic North American *T. officinale*; many are mixtures of diploid and triploid *Taraxacums*.
8. Haughton 1978, 104; Mitich 1989, 538; Stewart-Wade et al. 2002, 825; Sturtevant 1886, 5.
9. Stewart-Wade et al. 2002, 832.
10. Kalm 1937, 379.
11. The poems are "The Dandelion's pallid tube," "None who saw it ever told it," "Its little Ether Hood," and "I started Early—Took my Dog."
12. Jenkins 2015, 117.
13. Robbins 2012.
14. From the North Wind section of Henry Wadsworth Longfellow's "Hiawatha and Mudjekeewis."
15. Solbrig 1971, 687.
16. Darwin 1897, 551.
17. Troyer 2001, 291.
18. Davis 1979, 85.
19. Robbins 2012, 115.
20. Rüegg, Quadranti, and Zoschke 2007, 272.
21. Grube et al. 2011, 115.
22. Bernfeld 1946, 6–7.
23. Walt Whitman, "The First Dandelion," *New York Herald*, March 12, 1888, 4.
24. Grube et al. 2011, 25.
25. Rutherford and Deacon 1974, 251.
26. Rasmussen 2001, 299.
27. Mart 2015, 148.
28. Garabrant and Philbert 2002, 233; Loomis et al. 2015, 891; Perry 2008, 230.
29. Alavanja, Ross, and Bonner 2013, 122.
30. Cox and Surgan 2006, 1803.
31. Harris and Solomon 1992, 10.
32. Marcato, Souza, and Fontanetti 2017, 120.
33. Grover, Maybank, and Yoshida 1972, 320.
34. Alavanja, Ross, and Bonner 2013, 120.
35. Benfeito et al. 2014, 8; Ioannou, Fortun, and Tempest 2019, R15; Perry 2008, 233.
36. Alexander et al. 2017, 1.
37. Harlan and Wet 1965, 17.
38. Finlay 2009, 45.

39. Whaley and Bowen 1947, 11.
40. Beilen and Poirier 2007, 218.
41. Iaffaldano 2016, 1–10.
42. Zhang et al. 2017, 34.
43. The two species differ in *ploidy*, i.e., the number of copies of DNA in the genome. Dandelion is triploid (3n); TK is a diploid (2n) obligate outcrossing species. If crossed, the chromosomes would be unbalanced, and the ploidy of any progeny would be uncertain and likely infertile.

2. Florida Beggarweed

This chapter is dedicated to Dr. Ted Webster, my colleague and friend whose contribution to this work is sorely missed.

1. Lorenzi 2000, 36; Wilbur 1963, 152; Zimdahl 1989, 29.
2. Buchanan, Murray, and Hauser 1982, 219.
3. Byng et al. 2016, 17.
4. Burow et al. 2009, 107; Moretzsohn et al. 2004, 11.
5. Lorenzi 2000, 37; Schubert 1945, 430.
6. Hammons, Herman, and Stalker 2016, 3.
7. Jones 1885, 4; Smith 2002, 49.
8. Hammons, Herman, and Stalker 2016.
9. Simpson, Krapovickas, and Valls 2001, 79.
10. Andel et al. 2014, E5346; Murphy 2011, 29.
11. Berhaut 1975, 3:29; Brooks 1975, 29; Rodney 1970, 101.
12. French botanist Michel Adanson didn't mention them in writings about travels through Senegal and the Gambia in the mid-1700s. Swedish botanist Adam Afzelius wrote of "ground nuts" in Sierra Leone (~1790), but he was probably referring to native West African Bambara groundnuts (*Vigna subterranean*).
13. Hammons, Herman, and Stalker 2016, 5.
14. Carney and Rosomoff 2009, 65; Dow 1927, 90.
15. Cañizares-Esguerra and Breen 2013, 580; Kupperman 2012, 80, 96.
16. Carney and Rosomoff 2009, 89.
17. Crosby 2003, 208.
18. Carney and Rosomoff 2009, 154.
19. Hammons, Herman, and Stalker 2016, 8; Johnson 1964, 4.
20. Sloane 1707.
21. Sloane 1707, 2:185–86.
22. *Hedysarum* is one of several genera to which the species has been assigned.
23. Sloane 1707, 2:186.
24. Pickering 1879, 984.
25. Hooker et al. 1849, 122.
26. Candolle 1855, 783.
27. Swartz 1788.
28. Johnson 1964, 30.
29. Festinger 1983, 140.
30. Cumo 2015, 280.
31. Jefferson 1944, 29.
32. Clarke 1829, 451–54.
33. Bartram 1998, 153, 525.
34. Webster and Cardina 2004, 189.

35. Rogin 1991, 246.
36. Rogin 1991, 217.
37. Venables 2004, 157.
38. Bederman 1970, 72; Bonner 2009, 93–95.
39. Latham 1835, 8.
40. Rastogi, Pandey, and Rawat 2011, 286.
41. Tabuti, Lye, and Dhillion 2003, 31.
42. Brown 1938, 397.
43. Burrows and Wallace 1999, 867.
44. Chernow 2017, 465.
45. Brown 1938, 503.
46. Gould 2002, 38, 331.
47. Smith 2002, 23–24.
48. Spence 1902, 6.
49. Anonymous 1897, 5.
50. S. W. B. 1878.
51. Breese 1888.
52. Anonymous 1897.
53. Smith 1896, 16.
54. Hughes, Heath, and Metcalfe 1951.
55. Osband 1985, 627.
56. Cumo 2015, 286.
57. Cumo 2015, 281.
58. Jones 1885, 55; Kremer 2011, 97.
59. Timmons 1970, 295.
60. Buchanan, Murray, and Hauser 1982, 207.
61. Hauser and Buchanan 1977, 4–5.
62. Banks 1989, 566; Wilcut et al. 1989, 385.
63. Bockholts and Heidebrink 2012, 325–26.
64. Webster and Coble 1997, 308; Webster and Nichols 2012, 145.
65. Webster and Sosnoskie 2010, 73; Willingham et al. 2008, 74.
66. Ford 1978, 963; Tandyekkal 1997, 663; Tsai 2011, 474.

3. Velvetleaf

1. Burrows and Wallace 1999, 301.
2. Morison 1950, 8–10.
3. Sharma 1993, 274.
4. Vavílov 1992, 311.
5. Hu 1955, 31.
6. Henry 1891, 250.
7. Linnaeus 1753, 2:685.
8. Zimdahl 1989, 4.
9. Dodoens 1987.
10. Hymowitz and Harlan 1983, 371; Swingle 1945, 88.
11. Ewan and Ewan 1970, 108.
12. Gronovius and Clayton 1762, 79.
13. Miller 1754, 8.
14. Barton 1827, 91.
15. Barton 1827, 91.

16. Jefferson 1999, 53.
17. Muhlenberg 1793, 174.
18. Madison 1810.
19. Madison 1815.
20. Jensen 1969, 108.
21. Ratner 1972, 17.
22. American State Papers 1825.
23. Morison 1961, 254.
24. Darlington 1826, 77.
25. Darlington 1859, 65.
26. Allen 1870, 1.
27. Ellstrand 2019, 550.
28. Darlington 1859, 66.
29. Cardina and Sparrow 1997, 65.
30. Cardina and Norquay 1997, 90.
31. Weiner and Freckleton 2010, 173–90.
32. Howson 1862, 2–3.
33. Dodge 1880, 509.
34. Dodge 1880, 508.
35. New Jersey, State of 1881, 89.
36. Dodge 1880, 509.
37. Dodge 1894, 27.
38. New Jersey, State of 1881, 95–99.
39. New Jersey, State of 1881, 102.
40. Anonymous 1879, 2.
41. Mears and Charles 1898, 10.
42. Anonymous 1881, 4.
43. Dodge 1894, 28.
44. Bensky, Gamble, and Kaptchuk 2004, 300.
45. Hwang et al. 2013, 1918.
46. Kaufmann 1999, 456.
47. Gerhard 1857, 237–38.
48. Dies 1942, 18.
49. US Department of Agriculture 1942, 2–4.
50. Alachlor is no longer sold in the United States; it, along with atrazine, is illegal in Europe; in 2019 safeguards for atrazine were weakened, in spite of links to birth defects and cancer.
51. Liebman, Mohler, and Staver 2001, 11–13.
52. Kniss 2017, 5.
53. Kremer and Spencer 1989, 211.
54. Medović and Horváth 2012, 215.

4. Nutsedge

1. Sheahan, Barrett, and Goldvale 2017, 27.
2. Holm et al. 1977, 131.
3. Bendixen and Nandihalli 1987, 62.
4. Vries 1991, 29.
5. De Castro et al. 2015, 741.
6. Zohary and Hopf 2000, 2:158.

7. Negbi 1992, 65.
8. Negbi 1992, 62.
9. Schippers, Borg, and Bos 1995, 475.
10. Watts 2011, 2–3.
11. Tumbleson and Kommedahl 1961, 650.
12. Lapham and Drennan 1990, 125.
13. Thullen and Keeley 1979, 502.
14. Okoli et al. 1997, 111.
15. Baumann 1928, 290–91.
16. United Nations 2009.
17. Yuan et al. 2011, 264.
18. Jomo et al. 2016, 1–2.
19. Gouse et al. 2016, 27.
20. Sheahan, Barrett, and Goldvale 2017, 27.
21. Carrier 1923, 239.
22. De Castro et al. 2015, 740–41.
23. Romans 1775, 129.
24. Anonymous 1864; Anonymous 1906.
25. Gray 1862, 490.
26. Dewey 1895, 28.
27. Gray, Robinson, and Fernald 1906, 176.
28. Stoller 1981, 8.
29. Drost and Doll 1980, 229.
30. Tumbleson and Kommedahl 1961, 646.
31. Jing et al. 2016, 325.
32. Codina-Torrella, Guamis, and Trujillo 2015, 406.
33. Pascual et al. 2000, 442.
34. Chukwuma, Obioma, and Cristopher 2010, 710.
35. Gambo and Da'u 2014, 60.
36. Olabiyi, Oboh, and Adefegha 2017.
37. Allouh, Daradka, and Abu Ghaida 2015, 331.
38. Gunnell et al. 2007, 1235.
39. Yuan et al. 2011, 250.
40. Frick et al. 1979, 182.
41. Scheepens and Hoogerbrugge 1991, 245.
42. Coryndon et al. 1972.

5. Marestail

1. Dauer, Mortensen, and Humston 2006, 484.
2. Bruce and Kells 1990, 646.
3. Nandula et al. 2006, 901.
4. Nesom 1990, 231.
5. Zimdahl 1989, 22.
6. Hurt 1981.
7. Faulkner 1943, 170.
8. No-till, minimum-till, strip-till, and others are conservation tillage practices; I use the term no-till to represent any of these. Hornbeck 2012.
9. Derpsch et al. 2010, 8.
10. Marguiles 2012, 20.

11. Devine and Vail 2015, 2.

12. Chaganti and Culman 2017, 1–7.

13. Buchholtz et al. 1960, 718, 183.

14. Zimdahl 1989, 23–24.

15. Loux and Berry 1991, 464.

16. Bevan, Flavell, and Chilton 1983, 184.

17. *N*-(phosphonomethyl)glycine is an active ingredient in many herbicides marketed by several companies. Monsanto was acquired by Bayer in 2018.

18. EPSPS is the enzyme 5-enolpyruvylshikimate 3-phosphate synthase (EC 2.5.1.19).

19. Duke and Powles 2008, 319.

20. Bruce and Kells 1990, 645.

21. Padgette et al. 1995, 1452.

22. Davidson 1996, 174.

23. Dentzman 2018, 124.

24. Nachtigal 2001, 58.

25. Jasieniuk 1995, 26.

26. Bradshaw et al. 1997, 195.

27. The first global case of glyphosate resistance occurred in Australia, in a species of *Lolium*.

28. VanGessel 2001, 703.

29. Heap 2019.

30. Heap 2019.

31. Foresman and Glasgow 2008, 389.

32. Heap 2019.

33. Sammons and Gaines 2014.

34. Baucom 2016, 181–82.

35. Gressel 2002, 97.

36. Markus et al. 2018, 277.

37. Delye, Jasieniuk, and Le Corre 2013, 651; Sammons and Gaines 2014, 1373.

38. Dentzman 2018, 125.

39. Jordan 2002, 524.

40. Davidson 1996, 175.

41. Ortega et al. 1994, 598.

6. Pigweed

1. Valencius 2002, 109.

2. Gatewood and Whayne 1996.

3. Ward, Webster, and Steckel 2013, 12; Steckel 2007, 567.

4. Not to be confused with lambsquarters (*Chenopodium album*) or other species sometimes called "pigweed."

5. Sauer 1967, 12.

6. Sauer 1988, 102; Trucco and Tranel 2011, 12.

7. Sauer 1957, 25.

8. Waterhemp is also called "tall waterhemp" and "common waterhemp"; some consider them separate species although studies suggesting otherwise.

9. Sosnoskie et al. 2009, 404.

10. Trucco and Tranel 2011, 12.

11. Robbins, Crafts, and Raynor 1952, 121–28.

12. Heap 2019.

13. Kroma and Flora 2003, 26.

14. Horak and Peterson 1995, 192.

15. Gallagher and Cardina 1998a, 48; Gallagher and Cardina 1998b, 53.

16. Bradshaw et al. 1997, 195.

17. Koo et al. 2018, 1933.

18. Similar genetic modifications made by Dow allow soybean crops to tolerate 2,4-D. I tell the story of dicamba, but it is similar for 2,4-D.

19. Monsanto and BASF developed the low-volatile version of dicamba; DowDuPont introduced crops resistant to 2,4-D.

20. Koon 2017; Smith 2016.

21. Hutcheson 2016.

22. Koon 2017.

23. Bradley 2018.

24. Jones, Norsworthy, and Barber 2019, 41.

25. Hakim 2017; Peacock 2018.

26. Greene 2019.

27. Koon 2017.

28. Unglesbee 2019.

29. Peterson et al. 2019.

30. There is some botanical justice in this and other findings of evolved resistance in Kansas, the US state known for prohibiting the teaching of evolution in public schools.

31. Davis and Frisvold 2017, 2209.

32. Harvey 2018, 12.

33. Stepanek, Evertson, and Martens 2018.

34. Palhano, Norsworthy, and Barber 2018, 60; Price et al. 2016, 1.

35. "Clustered regularly interspaced short palindromic repeats," CRISPR, works in conjunction with the "cas9" protein to target specific places in the genome of an organism.

36. National Academies of Sciences, Engineering, and Medicine 2016.

37. Neve 2018, 2676.

38. Sammons and Gaines 2014, 1367.

39. Ribeiro et al. 2014, 207.

40. Sauer 1957, 24.

41. For a single species, the prize goes to *Lolium*, a grass.

7. Ragweed

1. Strictly speaking, "hay fever" pertains to grass allergy as opposed to "ragweed allergy"; I conflate them according to common usage.

2. Barnett and Steckel 2013, 547.

3. A third, not yet an international pest, is Cuman (or "western") ragweed (*A. psilostachya*), a perennial most common west of the Mississippi.

4. Martin et al. 2018, 336.

5. Abul-Fatih and Bazzaz 1979, 813.

6. Błoszyk et al. 1992, 1092.

7. A single seed is enclosed in a dry nutlike fruit called an "achene"; I use "seed" for convenience.

8. Martin, Zim, and Nelson 1961, 420.

9. Curtis and Lersten 1995, 34.

10. Fumanal, Charvel, and Bertagnolle 2007, 233.

11. Belmonte et al. 2000, 96.
12. Friedman and Barrett 2008, 1303.
13. Bordas-Le Floch et al. 2015, 2; Diethart, Sam, and Weber 2007, 164.
14. Gremillion 1996, 530.
15. Moreman 2003.
16. Wulf 2015, 58.
17. Whitney 1996, 228.
18. Darlington 1847, 80.
19. Whitney 1996, 271–74.
20. Blackmore et al. 2007, 483.
21. Bassett and Terasmae 1962, 147.
22. Blackley 1873, 75.
23. Beard 1881, vi.
24. Waite 1995, 194.
25. Mitman 2003, 610.
26. Waite 1995, 196.
27. Mitman 2003, 625.
28. Smith, Hamner, and Carlson 1946, 474.
29. Horning et al. 1941, 315.
30. Bullock et al. 2010, 137.
31. Walzer and Siegel 1956, 125.
32. Makra et al. 2015, 500; Stępalska et al. 2017, 749.
33. Essl et al. 2015, 1083.
34. Sato et al. 2017, 62.
35. Chauvel et al. 2006, 669.
36. Bullock et al. 2010, 34.
37. Oh et al. 2009, 56.
38. Bousquet et al. 2008, 1302.
39. Conquest 1986.
40. Kiss and Beres 2006; Reznik 2009, 2.
41. Makra et al. 2005, 57.
42. Taramarcaz et al. 2005, 538.
43. Bullock et al. 2010, 55; Galzina et al. 2010, 75.
44. Lim, Kim, and Park 2009, 506.
45. Bullock et al. 2010, 40.
46. Regnier et al. 2016, 362.
47. Regnier et al. 2008, 1627.
48. Van Horn et al. 2018, 1071.
49. Ding et al. 2008, 319.
50. Mayer 2018.
51. JinQing et al. 2017, 4120.
52. Chin and Spegele 2014.
53. Krstić et al. 2018, 435.
54. Yakkala et al. 2013, 62.
55. Hodgins and Rieseberg 2011, 2732.
56. Leiblein-Wild, Kaviani, and Tackenberg 2014, 744.
57. Simberloff 2006, 912.
58. Steffen et al. 2018, 8255.
59. Wayne et al. 2002, 280.

60. Ziska et al. 2003, tables 1 and 3.
61. Rogers et al. 2006, 868.
62. Ziska et al. 2011, 4249.
63. Pasqualini et al. 2011, 2830.
64. El Kelish et al. 2014.
65. Cunze, Leiblein, and Tackenberg 2013, 7.
66. Montagnani et al. 2017, 143.

8. Foxtail

1. Nurse et al. 2009, 383.
2. Alachlor is outlawed in Europe and no longer used in the United States.
3. Brutnell et al. 2010, 2540.
4. Benabdelmouna et al. 2001, 685.
5. Wang, Wendel, and Dekker 1995, 1037.
6. Callen 1976, 535.
7. Nurse et al. 2009, 383; Wang, Wendel, and Dekker 1995, 1031.
8. Kilpatrick 2014, 138–41.
9. Ferguson 1925.
10. Chase 1948, 14.
11. Fernald 1944, 57–58.
12. Pohl 1951, 505.
13. Peterson 1967, 252.
14. "ITIS Standard Report Page: Setaria Faberi" 2019.
15. Pohl 1951, 507.
16. Rominger 1962, 85.
17. Cardina, Regnier, and Harrison 1991, 186.
18. Dekker 2003, 85.
19. Dekker 2000, 411–23.
20. Thoreau 1962, 164.
21. Thoreau 2001.
22. Cardina et al. 2007, 455.
23. Heap 2019.
24. Kline 1990.
25. Zeisberger, Hulbert, and Schwarze 1910, 44.
26. Franklin 1786, 452.
27. Rominger 1962, 8.
28. Thoreau 1993.

References

Abul-Fatih, H. A., and F. A. Bazzaz. 1979. "The Biology of Ambrosia Trifida L. I. Influence of Species Removal on the Organization of the Plant Community." *New Phytologist* 83, no. 3: 813–16. https://doi.org/10.1111/j.1469-8137.1979.tb02313.x.

Alavanja, Michael C. R., Matthew K. Ross, and Matthew R. Bonner. 2013. "Increased Cancer Burden among Pesticide Applicators and Others Due to Pesticide Exposure." *CA: A Cancer Journal for Clinicians* 63, no. 2: 120–42. https://doi.org/10.3322/caac.21170.

Alexander, Melannie, Stella Koutros, Matthew R. Bonner, Kathryn Hughes Barry, Michael C. R. Alavanja, Gabriella Andreotti, Hyang-Min Byun et al. 2017. "Pesticide Use and LINE-1 Methylation among Male Private Pesticide Applicators in the Agricultural Health Study." *Environmental Epigenetics* 3, no. 2: 1–9. https://doi.org/10.1093/eep/dvx005.

Allen, J. A. 1870. "American Weeds." *Massachusetts Ploughman*. August 27. America's Historical Newspapers.

Allouh, M. Z., Daradka, H. M., and Abu Ghaida, J. H. 2015. "Influence of Cyperus Esculentus Tubers (Tiger Nut) on Male Rat Copulatory Behavior." *BMC Complementary and Alternative Medicine* 15: 331–38. https://doi.org/10.1186/s12906-015-0851-9.

American State Papers. 1825. Naval Affairs, 18th Congress, 2nd Session. "On the Use of American Canvas, Cables and Cordage in the Navy." Washington, DC: Congress of the United States, January 10. https://memory.loc.gov/cgi-bin/query/S?ammem/ hlaw:@filreq(@band(+@1(Canvas,+Cables+and+Cordage))+@field(COLLID+llsp)).

Andel, Tinde R. van, Charlotte I.E.A. van't Klooster, Diana Quiroz, Alexandra M. Towns, Sofie Ruysschaert, and Margot van den Berg. 2014. "Local Plant Names Reveal That Enslaved Africans Recognized Substantial Parts of the New World Flora." *Proceedings of the National Academy of Sciences* 111, no. 50 (December 16): E5346–53. https://doi.org/10.1073/pnas.1418836111.

Anonymous. 1864. "The Chufa for Hogs." *Edgefield Advertiser*. February 10. http:// chroniclingamerica.loc.gov/lccn/sn84026897/1864-02-10/ed-1/seq-2/.

Anonymous. 1879. "A Valuable Weed." *Daily Inter Ocean* 8, no. 106: 2.

Anonymous. 1881. "Laws of New Jersey Chapter LXXV." *Trenton State Gazette Legislative Newspaper* 35, no. 67 (March 19): 4.

Anonymous. 1897. "Echoes from the Fair." *News and Observer*. October 24.

Anonymous. 1906. "The Chufa or Ground Almond." *The Florida Agriculturist*. June 13, Image 2. https://chroniclingamerica.loc.gov/lccn/sn96027724/1906-06-13/ed-1/.

Baker, Herbert G. 1965. "Characteristics and Modes of Origin of Weeds." In *The Genetics of Colonizing Species*, edited by H. G. Baker and G. L. Stebbins, 147–72. New York: Academic Press.

Banks, P.A. 1989. "Herbicide Replacements for Dinoseb in Peanuts." *Research Report—University of Georgia, College of Agriculture, Agricultural Experiment Stations* (April), 566–73.

Barnett, Kelly A., and Lawrence E. Steckel. 2013. "Giant Ragweed (Ambrosia Trifida) Competition in Cotton." *Weed Science* 61, no. 4: 543–48. https://doi.org/10.1614/ WS-D-12-00169.1.

Barton, Benjamin Smith. 1827. *Elements of Botany: Outlines of the Natural History of Vegetables*. 3rd ed. Philadelphia: Robert Desilver. https://www.biodiversitylibrary. org/item/84262.

Bartram, William, and Francis Harper. 1998. *The Travels of William Bartram Edited with Commentary by Francis Harper*. Athens: University of Georgia Press.

Bassett, I.J., and J. Terasmae. 1962. "Ragweeds, Ambrosia Species, in Canada and Their History in Postglacial Time." *Canadian Journal of Botany* 40, no. 1: 141–50. https://doi.org/10.1139/b62-015.

Baucom, Regina S. 2016. "The Remarkable Repeated Evolution of Herbicide Resistance." *American Journal of Botany* 103, no. 2: 181–83. https://doi.org/10.3732/ ajb.1500510.

Baumann, Hermann. 1928. "The Division of Work According to Sex in African Hoe Culture." *Africa: Journal of the International African Institute* 1, no. 3: 289–319. https://doi.org/10.2307/1155633.

Beard, George Miller. 1881. *American Nervousness, Its Causes and Consequences: A Supplement to Nervous Exhaustion (Neurasthenia)*. New York: G.P. Putnam's Sons.

Bederman, S.H. 1970. "Recent Changes in Agrarian Land Use in Georgia." *Southeastern Geographer* 10, no. 2: 72–82.

Beilen, Jan B. van, and Yves Poirier. 2007. "Guayule and Russian Dandelion as Alternative Sources of Natural Rubber." *Critical Reviews in Biotechnology* 27, no. 4: 217–31. https://doi.org/10.1080/07388550701775927.

Belmonte, Jordina, Mercè Vendrell, Joan M. Roure, Josep Vidal, Jaume Botey, and Àlvar Cadahía. 2000. "Levels of Ambrosia Pollen in the Atmospheric Spectra of Catalan Aerobiological Stations." *Aerobiologia* 16, no. 1: 93–99. https://doi.org/10.1023/A:1007649427549.

Benabdelmouna, A., Y. Shi, M. Abirached-Darmency, and H. Darmency. 2001. "Genomic in Situ Hybridization (GISH) Discriminates between the A and the B Genomes in Diploid and Tetraploid Setaria Species." *Genome* 44, no. 4: 685–90. https://doi.org/10.1139/gen-44-4-685.

Bendixen, Leo E., and U.B. Nandihalli. 1987. "Worldwide Distribution of Purple and Yellow Nutsedge (Cyperus Rotundus and C. Esculentus)." *Weed Technology* 1, no. 1: 61–65. http://www.jstor.org/stable/3986985.

Benfeito, S., T. Silva, J. Garrido, P. Andrade, M. Sottomayor, F. Borges, F., and E. Garrido. 2014. "Effects of Chlorophenoxy Herbicides and Their Main Transformation Products on DNA Damage and Acetylcholinesterase Activity." *BioMed Research International* 2014: 1–10. https://doi.org/10.1155/2014/709036.

Bensky, Dan, Andrew Gamble, and Ted J. Kaptchuk. 2004. *Chinese Herbal Medicine: Materia Medica*. 3rd ed. Seattle, WA: Eastland Press.

Berhaut, Jean. 1975. *Flore Illustrée Du Sénégal: Dicotylédones*. Vol. 3. Dakar: Gouvernement du Sénégal.

Bernfeld, Siegfried. 1946. "An Unknown Autobiographical Fragment by Freud." *American Imago* 4, no. 1: 3–19. http://search.proquest.com/docview/1289640320/citation/CD0067DF87CC4800PQ/18.

Bevan, Michael W., Richard B. Flavell, and Mary-Dell Chilton. 1983. "A Chimaeric Antibiotic Resistance Gene as a Selectable Marker for Plant Cell Transformation." *Nature* 304, no. 5922: 184–87. https://doi.org/10.1038/304184a0.

Blackley, Charles Harrison. 1873. *Experimental Researches on the Causes and Nature of Catarrhus Æstivus (Hay-Fever Or Hay-Asthma)*. London: Baillière, Tindall & Cox.

Blackmore, Stephen, Alexandra H. Wortley, John J. Skvarla, and John R. Rowley. 2007. "Pollen Wall Development in Flowering Plants." *New Phytologist* 174, no. 3: 483–98. https://doi.org/10.1111/j.1469-8137.2007.02060.x.

Blatchley, Willis Stanley. 1912. *The Indiana Weed Book*. Indianapolis: Nature Publishing Company.

Błoszyk, Elżbieta, Urszula Rychłewska, Beata Szczepanska, Miloš Budĕšínský, Bohdan Drożdż, and Miroslav Holub. 1992. "Sesquiterpene Lactones of Ambrosia Artemisiifolia L. and Ambrosia Trifida L. Species." *Collection of Czechoslovak Chemical Communications* 57, no. 5: 1092–102. https://doi.org/10.1135/cccc19921092.

Bockholts, P., and I. Heidebrink. 2012. *Chemical Spills and Emergency Management at Sea: Proceedings of the First International Conference on "Chemical Spills and Emergency Management at Sea."* Amsterdam, The Netherlands, November 15–18, 1988. Springer Science and Business Media.

Bonner, James C. 2009. *History of Georgia Agriculture, 1732–1860*. University of Georgia Press.

Bordas-Le Floch, Véronique, Maxime Le Mignon, Julien Bouley, Rachel Groeme, Karine Jain, Véronique Baron-Bodo, Emmanuel Nony, Laurent Mascarell, and Philippe Moingeon. 2015. "Identification of Novel Short Ragweed Pollen Allergens

Using Combined Transcriptomic and Immunoproteomic Approaches." *PLOS ONE*, edited by Bernhard Ryffel 10, no. 8: e0136258. https://doi.org/10.1371/journal. pone.0136258.

Bousquet, P.-J., B. Leynaert, F. Neukirch, J. Sunyer, C.M. Janson, J. Anto, D. Jarvis, and P. Burney. 2008. "Geographical Distribution of Atopic Rhinitis in the European Community Respiratory Health Survey." *Allergy* 63, no. 10: 1301–9. https://doi. org/10.1111/j.1398-9995.2008.01824.x.

Bradley, K. 2018. "Dicamba Injury Mostly Confined to Specialty Crops, Ornamentals, and Trees so Far." *Integrated Pest Management—University of Missouri* (blog). June 6. https://ipm.missouri.edu/IPCM/2018/6/dicambaInjuryConfined.

Bradshaw, L.D., S. R. Padgette, S. L. Kimball, and B. H. Wells. 1997. "Perspectives on Glyphosate Resistance." *Weed Technology* 11: 189–98. https://doi.org/www.jstor. org/stable/3988252.

Breese, W.W. 1888. "Florida as It Is." *Daily Inter Ocean*. August 13.

Brooks, George E. 1975. "Peanuts and Colonialism: Consequences of the Commercial- ization of Peanuts in West Africa, 1830–70." *Journal of African History* 16, no. 1: 29–54. http://www.jstor.org/stable/181097.

Brown, Harry Bates. 1938. *Cotton: History, Species, Varieties, Morphology, Breeding, Culture, Diseases, Marketing, and Uses*. London: McGraw-Hill.

Bruce, J.A., and J. J. Kells. 1990. "Horseweed (Conyza Canadensis) Control in No- Tillage Soybeans (Glycine Max) with Preplant and Preemergence Herbicides." *Weed Technology* 4, no. 3: 642–47. https://doi.org/10.1017/S0890037X00026130.

Brutnell, Thomas P., Lin Wang, Kerry Swartwood, Alexander Goldschmidt, David Jackson, Xin-Guang Zhu, Elizabeth Kellogg, and Joyce Van Eck. 2010. "Setaria Viri- dis: A Model for C4 Photosynthesis." *Plant Cell* 22, no. 8: 2537–44. https://doi. org/10.1105/tpc.110.075309.

Buchanan, G.A., D. S. Murray, and E.W. Hauser. 1982. "Weeds and Their Control in Peanuts." In *Peanut Science and Technology*, edited by H.E. Pattee and C.T. Young, 206–49. Stillwater, OK: American Peanut Research and Education Society. https:// apresinc.com/wp-content/uploads/2015/12/PST-Chapter-8.pdf.

Buchholtz, K.P., B. H. Grigsby, O. C. Lee, F. W. Slife, J. C. Willard, and N. J. Volk. 1960. *Weeds of the North Central States*. Vol. 718. North Central Regional Publica- tion 36. Urbana, IL: University of Illinois Agricultural Experiment Station.

Bullock, James M., D. Chapman, S. Schafer, R. M. Girardello, T. Haynes, and S. Beal et al. 2010. "Assessing and Controlling the Spread and the Effects of Common Rag- weed in Europe." Final. Wallingford, UK: Natural Environment Research Council, UK. https://circabc. europa. eu/sd/d/d1ad57e8-327c-4fdd-b908-dadd5b859eff/Final_ Final_Report. pdf.

Burow, Mark D., Charles E. Simpson, Michael W. Faries, James L. Starr, and Andrew H. Paterson. 2009. "Molecular Biogeographic Study of Recently Described B- and A-Genome Arachis Species, Also Providing New Insights into the Origins of Cul- tivated Peanut." *Genome* 52, no. 2 (January 15): 107–19. https://doi.org/10.1139/ G08-094.

Burrows, Edwin G., and Mike L. Wallace. 1999. *Gotham: A History of New York City to 1898*. New York: Oxford University Press.

Byng, James W., Mark W. Chase, Maarten J. M. Christenhusz, Michael F. Fay, Walter S. Judd, David J. Mabberley, Alexander N. Sennikov, Douglas E. Soltis, Pamela S. Soltis, and Peter F. Stevens. 2016. "An Update of the Angiosperm Phylogeny Group Classification for the Orders and Families of Flowering Plants: APG IV." *Botanical Journal of the Linnean Society* 181, no. 1: 1–20.

Callen, E. O. 1976. "The First New World Cereal." *American Antiquity* 32, no. 4: 535–38. https://www.jstor.org/stable/2694082.

Candolle, Alphonse de. 1855. *Géographie botanique raisonnée: ou, Exposition des faits principaux et des lois concernant la distribution géographique des plantes de l'epoque actuelle.* V. Masson.

Cañizares-Esguerra, Jorge, and Benjamin Breen. 2013. "Hybrid Atlantics: Future Directions for the History of the Atlantic World." *History Compass* 11, no. 8 (August 1): 597–609. https://doi.org/10.1111/hic3.12051.

Cardina, J., C. P. Herms, D. A. Herms, and F. Forcella. 2007. "Evaluating Phenological Indicators for Predicting Giant Foxtail (Setaria Faberi) Emergence." *Weed Science* 55: 455–64. https://doi.org/10.1614/WS-07-005.1.

Cardina, John, and Heather M. Norquay. 1997. "Seed Production and Seedbank Dynamics in Subthreshold Velvetleaf (Abutilon Theophrasti) Populations." *Weed Science* 1: 85–90. www.jstor.org/stable/4045717.

Cardina, John, Emilie Regnier, and Kent Harrison. 1991. "Long-Term Tillage Effects on Seed Banks in Three Ohio Soils." *Weed Science* 39, no. 2: 186–94. http://www.jstor.org/stable/4044914.

Cardina, John, and Denise H. Sparrow. 1997. "Temporal Changes in Velvetleaf (Abutilon Theophrasti) Seed Dormancy." *Weed Science* 1: 61–66. http://www.jstor.org/stable/4045714.

Carney, J. A. and R. N. Rosomoff. 2009. *In the Shadow of Slavery: Africa's Botanical Legacy in the Atlantic World.* Berkeley: University of California Press.

Carrier, Lyman. 1923. *The Beginnings of Agriculture in America.* New York: McGraw-Hill.

Chaganti, Vijayasatya Nagendra, and Steven W. Culman. 2017. "Historical Perspective of Soil Balancing Theory and Identifying Knowledge Gaps: A Review." *Crop, Forage and Turfgrass Management* 3, no. 1: 1–7. https://doi.org/10.2134/cftm2016.10.0072.

Chase, Agnes. 1948. "The Meek That Inherit the Earth." In *Grass, The Yearbook of Agriculture,* edited by A. Stefferud, 8–15. Washington, DC: U.S. Government Printing Office.

Chauvel, Bruno, Fabrice Dessaint, Catherine Cardinal-Legrand, and François Bretagnolle. 2006. "The Historical Spread of Ambrosia Artemisiifolia L. in France from Herbarium Records." *Journal of Biogeography* 33, no. 4: 665–73. https://doi.org/10.1111/j.1365-2699.2005.01401.x.

Chernow, Ron. 2017. *Grant.* New York: Penguin Press.

Chin, Josh, and Brian Spegele. 2014. "China Details Vast Extent of Soil Pollution." *Wall Street Journal,* World section. https://www.wsj.com/articles/china-says-nearly-one-fifth-of-its-arable-land-is-contaminated-by-pollution-1397726425.

Chukwuma, E. R., N. Obioma, and O. I. Cristopher. 2010. "The Phytochemical Composition and Some Biochemical Effects of Nigerian Tigernut (Cyperus Esculentus L.) Tuber." *Pakistan Journal of Nutrition* 9, no. 7: 709–15.

Clarke, George F. 1829. "On Green Crops buried as Manure." In *Southern Agriculturist and Register of Rural Affair Volume 2*, edited by J. D. Legare, October: 451–454. https://books.google.com/books?id=WfyOHMSoTnIC 451-454.

Codina-Torrella, Idoia, Buenaventura Guamis, and Antonio J. Trujillo. 2015. "Characterization and Comparison of Tiger Nuts (Cyperus Esculentus L.) from Different Geographical Origin." *Industrial Crops and Products* 65: 406–14. https://doi.org/10.1016/j.indcrop.2014.11.007.

Conquest, Robert. 1986. *The Harvest of Sorrow: Soviet Collectivization and the Terror-Famine*. New York: Oxford University Press.

Coryndon, Shirley C., A. W. Gentry, John M. Harris, D. A. Hooijer, Vincent J. Maglio, and F. Clark Howell. 1972. "Mammalian Remains from the Isimila Prehistoric Site, Tanzania." *Nature* 237, no. 5353: 292. https://doi.org/10.1038/237292a0.

Cox, Caroline, and Michael Surgan. 2006. "Unidentified Inert Ingredients in Pesticides: Implications for Human and Environmental Health." *Environmental Health Perspectives* 114, no. 12: 1803–6. https://doi.org/10.1289/ehp.9374.

Crosby, Alfred W. 2003. *The Columbian Exchange: Biological and Cultural Consequences of 1492*. Westport, CT: Greenwood.

Cumo, Christopher Martin. 2015. *Foods That Changed History: How Foods Shaped Civilization from the Ancient World to the Present*. Santa Barbara, CA: ABC-CLIO.

Cunze, S., M. C. Leiblein, and O. Tackenberg. 2013. "Range Expansion of Ambrosia Artemisiifolia in Europe Is Promoted by Climate Change." *ISRN Ecology* Article ID 610126: 1–9. http://dx.doi.org/10.1155/2013/610126.

Curtis, John D., and Nels R. Lersten. 1995. "Anatomical Aspects of Pollen Release from Staminate Flowers of Ambrosia Trifida (Asteraceae)." *International Journal of Plant Sciences* 156, no. 1: 29–36. https://doi.org/10.1086/297225.

Darlington, William. 1826. *Florula Cestrica, an Essay*. Westchester, PA: S. Siegfried. https://doi.org/10.5962/bhl.title.24908.

Darlington, William. 1847. *Agricultural Botany: An Enumeration and Description of Useful Plants and Weeds Which Merit the Notice or Require the Attention of American Agriculturists*. Philadelphia: J. W. Moore.

Darlington, William. 1859. *American Weeds and Useful Plants*. 2nd ed. New York: A. O. Moore. http://hdl.handle.net/2027/chi.086350831.

Darwin, Charles. 1897. *The Power of Movement in Plants*. New York: Appleton.

Dauer, Joseph T., David A. Mortensen, and Robert Humston. 2006. "Controlled Experiments to Predict Horseweed (Conyza Canadensis) Dispersal Distances." *Weed Science* 54, no. 3: 484–89. https://doi.org/10.1614/WS-05-017R3.1.

Davidson, O. G. 1996. *Broken Heartland: The Rise of America's Rural Ghetto*. 2nd ed. Iowa City: University of Iowa Press.

Davis, Adam S, and George B. Frisvold. 2017. "Are Herbicides a Once in a Century Method of Weed Control?" *Pest Management Science* 73, no. 11: 2209–20. https://doi.org/10.1002/ps.4643.

Davis, Donald E. 1979. "Herbicides in Peace and War." *BioScience* 29, no. 2 (February 1): 84–94. https://doi.org/10.2307/1307743.

De Castro, Olga, Roberta Gargiulo, Emanuele Del Guacchio, Paolo Caputo, and Paolo De Luca. 2015. "A Molecular Survey Concerning the Origin of Cyperus Esculentus

(Cyperaceae, Poales): Two Sides of the Same Coin (Weed vs. Crop)." *Annals of Botany* 115, no. 5: 733–45. https://doi.org/10.1093/aob/mcv001.

De Wet, J.M.J., and J.R. Harlan. 1975. "Weeds and Domesticates: Evolution in the Man-Made Habitat." *Economic Botany* 29, no. 2: 99–108. https://doi.org/10.1007/BF02863309.

Dekker, Jack. 2000. "Emergent Weedy Foxtail (Setaria Spp.) Seed Germinability Behaviour." In *Seed Biology: Advances and Applications: Proceedings of the Sixth International Workshop on Seeds, Mérida, México*, edited by Michael Black, Kent J. Bradford, and Jorge Vázquez-Ramos, 411–23. Wallingford, UK: CABI.

Dekker, Jack. 2003. "The Foxtail (Setaria) Species-Group." *Weed Science* 51, no. 5: 641–56. https://doi.org/10.1614/P2002-IR.

Delye, C., M. Jasieniuk, and V. Le Corre. 2013. "Deciphering the Evolution of Herbicide Resistance in Weeds." *Trends in Genetics* 29, no. 11: 649–59. http://dx.doi.org/10.1016/j.tig.2013.06.001.

Dentzman, K. 2018. "'I Would Say That Might Be All It Is, Is Hope': The Framing of Herbicide Resistance and How Farmers Explain Their Faith in Herbicides." *Journal of Rural Studies* 57: 118–27. https://doi.org/10.1016/j.jrurstud.2017.12.010.

Derpsch, Rolf, Theodor Friedrich, Amir Kassam, and Hongwen Li. 2010. "Current Status of Adoption of No-till Farming in the World and Some of Its Main Benefits." *International Journal of Agricultural and Biological Engineering* 3, no. 1: 1–25. https://doi.org/10.25165/ijabe.v3i1.223.

Devine, Jenny Barker, and David D. Vail. 2015. "Sustaining the Conversation: The Farm Crisis and the Midwest." *Middle West Review* 2, no. 1: 1–9. https://doi.org/10.1353/mwr.2015.0044.

Dewey, L.H. 1895. "Weeds and How to Kill Them." U.S. Dept. of Agriculture. Farmers' Bulletin. Washington, DC: U.S. Department of Agriculture. https://archive.org/details/CAT87203839/page/n1.

Dies, Edward Jerome. 1942. *Soybeans: Gold from the Soil,*. New York. http://hdl.handle.net/2027/mdp.39015063890720.

Diethart, Bernadette, Saskia Sam, and Martina Weber. 2007. "Walls of Allergenic Pollen: Special Reference to the Endexine." *Grana* 46, no. 3: 164–75. https://doi.org/10.1080/00173130701472181.

Ding, Jianqing, Richard N. Mack, Ping Lu, Mingxun Ren, and Hongwen Huang. 2008. "China's Booming Economy Is Sparking and Accelerating Biological Invasions." *BioScience* 58, no. 4: 317–24. https://doi.org/10.1641/B580407.

Dodge, C.R. 1894. *A Report on the Uncultivated Bast Fibers of the United States: Including the History of Previous Experiments with the Plants of Fibers, and Brief Statements Relating to the Allied Species That Are Produced Commercially in the Old World*. Fiber Investigations Report. Washington, DC: U.S. Government Printing Office. https://books.google.com/books?id=rmbMwAEACAAJ.

Dodge, Charles Richards. 1880. "Vegetable Fibers." Report of the Commissioner of Agriculture. Washington, DC: U.S. Department of Agriculture. https://books.google.com/books?id=CkFDAQAAMAAJ.

Dodoens, Rembert. 1987. *Cruydeboeck*. 1563; reprint Westport, CT: Meckler/Chadwyck-Healey.

Dow, George Francis. 1927. *Slave Ships and Slaving*. Kennikat Press.

Drost, Dirk C., and Jerry D. Doll. 1980. "The Allelopathic Effect of Yellow Nutsedge (Cyperus Esculentus) on Corn (Zea Mays) and Soybeans (Glycine Max)." *Weed Science* 28, no. 2: 229–33. https://doi.org/10.1017/S004317450005517X.

Duke, Stephen O., and Stephen B. Powles. 2008. "Glyphosate: A Once-in-a-Century Herbicide." *Pest Management Science* 64, no. 4: 319–25. https://doi.org/10.1002/ps.1518.

El Kelish, Amr, Feng Zhao, Werner Heller, Jörg Durner, J Winkler, Heidrun Behrendt, Claudia Traidl-Hoffmann et al. 2014. "Ragweed (Ambrosia Artemisiifolia) Pollen Allergenicity: SuperSAGE Transcriptomic Analysis upon Elevated CO2 and Drought Stress." *BMC Plant Biology* 14, no. 1: 176. https://doi.org/10.1186/1471-2229-14-176.

Ellstrand, Norman C. 2019. "The Evolution of Crops That Do Not Need Us Anymore." *New Phytologist* 224, no. 2: 550–51. https://doi.org/10.1111/nph.16145.

Emerson, Ralph Waldo. 1878. *Fortune of the Republic*. Boston: Houghton, Osgood. https://catalog.hathitrust.org/Record/006507827.

Essl, Franz, Krisztina Biró, Dietmar Brandes, Olivier Broennimann, James M. Bullock, Daniel S. Chapman, Bruno Chauvel et al. 2015. "Biological Flora of the British Isles: Ambrosia Artemisiifolia." *Journal of Ecology* 103, no. 4: 1069–98. https://doi.org/10.1111/1365-2745.12424.

Ewan, Joseph, and Nesta Dunn Ewan. 1970. *John Banister and His Natural History of Virginia, 1678–1692*. Champaign: University of Illinois Press.

Faulkner, Edward H. 1943. *Plowman's Folly*. Norman: University of Oklahoma Press.

Ferguson, W. C. 1925. "Chaetochloa Italica." C. V. Starr Virtual Herbarium. New York Botanical Garden. http://sweetgum.nybg.org/science/vh/specimen-details/?irn=2016269.

Fernald, M. L. 1944. "Setaria Faberii in Eastern America." *Rhodora* 46, no. 542: 57–58. https://www-jstor-org.proxy.lib.ohio-state.edu/stable/23302297.

Festinger, Leon. 1983. *The Human Legacy*. New York: Columbia University Press.

Finlay, Mark R. 2009. *Growing American Rubber: Strategic Plants and the Politics of National Security*. New Brunswick, NJ: Rutgers University Press.

Ford, C. W. 1978. "In Vitro Digestibility and Chemical Composition of Three Tropical Pasture Legumes, Desmodium Intortum Cv. Greenleaf, D. Tortuosum and Macroptilium Atropurpureum Cv. Siratro." *Australian Journal of Agricultural Research* 29, no. 5: 963–74.

Foresman, Chuck, and Les Glasgow. 2008. "US Grower Perceptions and Experiences with Glyphosate-Resistant Weeds." *Pest Management Science* 64: 388–91. https://doi-org.proxy.lib.ohio-state.edu/10.1002/ps.1535.

Franklin, Benjamin. 1786. "Remarks Concerning the Savages of North America." *New London Magazine* 2, no. 16 (September): 452–55. https://search.proquest.com/openview/e90d798d96afb2ba/1?pq-origsite=gscholar&cbl=2346.

Frick, K. E., R. D. Williams Jr., P. C. Quimby, and R. F. Wilson. 1979. "Comparative Biocontrol of Purple Nutsedge (Cyperus Rotundus) and Yellow Nutsedge (C. Esculentus) with Bactra Verutana under Greenhouse Conditions." *Weed Science*, 27, no. 2: 178–83. https://www.jstor.org/stable/4042999.

Friedman, Jannice, and Spencer C.H. Barrett. 2008. "High Outcrossing in the Annual Colonizing Species Ambrosia Artemisiifolia (Asteraceae)." *Annals of Botany* 101, no. 9: 1303–9. https://doi.org/10.1093/aob/mcn039.

Fumanal, B., B. Charvel, and F. Bertagnolle. 2007. "Estimation of Pollen and Seed Production of Common Ragweed in France." *Annals of Agricultural and Environmental Medicine* 14, no. 2: 233–36.

Gallagher, R.S., and J. Cardina. 1998a. "Phytochrome-Mediated Amaranthus Germination I: Effect of Seed Burial and Germination Temperature." *Weed Science* 46: 48–52. www.jstor.org/stable/4046007.

Gallagher, R.S., and J. Cardina. 1998b. "Phytochrome-Mediated Amaranthus Germination II: Development of Very Low Fluence Sensitivity." *Weed Science* 46: 53–58. https://doi.org/10.1017/S0043174500090160.

Galzina, Natalija, Klara Baric, Maja Ščepanovic, Matija Gorsic, and Zvonimir Ostojic. 2010. "Distribution of Invasive Weed Ambrosia Artemisiifolia L. in Croatia." *Agriculturae Conspectus Scientifi Cus* 75, no. 2: 75–81. https://hrcak.srce.hr/62460.

Gambo, A., and A. Da'u. 2014. "Tiger Nut (Cyperus Esculentus): Composition, Products, Uses and Health Benefits—a Review." *Bayero Journal of Pure and Applied Sciences* 7, no. 1: 56–61. http://dx.doi.org/10.4314/bajopas.v7i1.11.

Garabrant, David H., and Martin A. Philbert. 2002. "Review of 2,4-Dichlorophenoxyacetic Acid (2,4-D) Epidemiology and Toxicology." *Critical Reviews in Toxicology* 32, no. 4 (January 1): 233–57. https://doi.org/10.1080/20024091064237.

Gatewood Jr., B., and M. Whayne. 1996. *The Arkansas Delta: Land of Paradox.* Fayetteville: University of Arkansas Press. https://muse.jhu.edu/.

Gerhard, Frederick. 1857. *Illinois as It Is: Its History, Geography, Statistics, Constitution, Laws, Government, Finances, Climate, Soil, Plants, Animals, State of Health . . .* Chicago: Keen and Lee. http://galenet.galegroup.com/servlet/Sabin?af=RN&ae=CY10 7814412&srchtp=a&ste=14.

Gould, William B. 2002. *Diary of a Contraband: The Civil War Passage of a Black Sailor.* Stanford, CA: Stanford University Press. http://goulddiary.stanford.edu/.

Gouse, Marnus, Debdatta Sengupta, Patricia Zambrano, and José Falck Zepeda. 2016. "Genetically Modified Maize: Less Drudgery for Her, More Maize for Him? Evidence from Smallholder Maize Farmers in South Africa." *World Development* 83: 27–38. https://doi.org/10.1016/j.worlddev.2016.03.008.

Gray, Asa. 1862. *Manual of the Botany of the Northern United States.* 3rd ed. Chicago: Ivison, Phinney. https://www.biodiversitylibrary.org/item/63069.

Gray, Asa, Benjamin Lincoln Robinson, and Merritt Lyndon Fernald. 1906. *Gray's New Manual of Botany.* 7th ed. New York: American Book Company.

Greene, David. 2019. "Is Fear Driving Sales of Monsanto's Dicamba-Proof Soybeans?" Radio transcript. *The Salt* (blog). February 7. https://www.npr.org/templates/tran script/transcript.php?storyId=691979417.

Gremillion, Kristen J. 1996. "Early Agricultural Diet in Eastern North America: Evidence from Two Kentucky Rockshelters." *American Antiquity* 61, no. 3: 520–36. https://doi.org/10.2307/281838.

Gressel, Jonathan. 2002. *Molecular Biology of Weed Control.* Boca Raton: CRC Press.

Gronovius, Johannes Fredericus, and John Clayton. 1762. *Flora Virginica*. Lugduni Batavorum. https://www.biodiversitylibrary.org/bibliography/60264.

Grover, R., J. Maybank, and K. Yoshida. 1972. "Droplet and Vapor Drift from Butyl Ester and Dimethylamine Salt of 2,4-D." *Weed Science* 20, no. 4: 320–24. https://doi.org/10.1017/S004317450003575X.

Grube, Arthur, David Donaldson, Timothy Kiely, and La Wu. 2011. "Pesticide Industry Sales and Usage." Biological and Economic Analysis Division. Washington, DC: U.S. Environmental Protection Agency.

Gunnell, D., R. Fernando, M. Hewagama, F. Koradsen, and M. Eddleston. 2007. "The Impact of Pesticide Regulations on Suicide in Sri Lanka." *International Journal of Epidemiology* 36, no. 6: 1235–42. https://doi.org/10.1093/ije/dym164.

Hakim, D. 2017. "Seeds, Weeds and Divided Farmers." *New York Times*. September 22, sec. B. https://www.nytimes.com/2017/09/21/business/monsanto-dicamba-weed-killer.html.

Hammons, Ray O., Danielle Herman, and H. Thomas Stalker. 2016. "Origin and Early History of the Peanut." In *Peanuts*, edited by H. Thomas Stalker and Richard F. Wilson, 1–26. Amsterdam: Elsevier.

Harari, Yuval N. 2015. *Sapiens: A Brief History of Humankind*. London: Vintage.

Harlan, Jack R., and J.M.J. de Wet. 1965. "Some Thoughts about Weeds." *Economic Botany* 19, no. 1: 16–24. https://doi.org/10.1007/BF02971181.

Harris, S.A., and K.R. Solomon. 1992. "Human Exposure to 2,4-D Following Controlled Activities on Recently Sprayed Turf." *Journal of Environmental Science and Health, Part B* 27, no. 1: 9–22. https://doi.org/10.1080/03601239209372764.

Harvey, James S., Jr. 2018. *Ethical Tensions from New Technology: The Case of Agricultural Biotechnology*. Wallingford, UK: CABI.

Haughton, C.S. 1978. *Green Immigrants: The Plants That Transformed America*. New York: Harcourt Brace Jovanovich.

Hauser, E. W., and G. A. Buchanan. 1977. "Control of Broadleaf Weeds in Peanuts with Dinoseb." *Georgia Agricultural Research*, no. 2: 4–7.

Heap, I. 2019. "The International Survey of Herbicide Resistant Weeds." www.weedscience.org.

Henry, Augustine. 1891. "Memorandum on the Jute and Hemp of China." *Kew Gardens Bulletin of Miscellaneous Information* 218, no. 58–59: 248–56.

Hodgins, K.A., and L. Rieseberg. 2011. "Genetic Differentiation in Life-History Traits of Introduced and Native Common Ragweed (Ambrosia Artemisiifolia) Populations." *Journal of Evolutionary Biology* 24, no. 12: 2731–49. https://doi.org/10.1111/j.1420-9101.2011.02404.x.

Holm, L.G., D. L. Plucknett, J. V. Pancho, and J. P. Herberger. 1977. *The World's Worst Weeds: Distribution and Biology*. Honolulu: The University Press of Hawaii.

Hooker, William Jackson, George Bentham, Joseph Dalton Hooker, Philip Barker Webb, and Theodore Vogel. 1849. *Niger Flora; or, An Enumeration of the Plants of Western Tropical Africa*. London: H. Baillière. http//catalog.hathitrust.org/Record/001494619.

Horak, Michael J., and Dallas E. Peterson. 1995. "Biotypes of Palmer Amaranth (Amaranthus Palmeri) and Common Waterhemp (Amaranthus Rudis) Are Resistant to Imazethapyr and Thifensulfuron." *Weed Technology* 9: 192–95. https://doi. org/10.1017/S0890037X00023174.

Hornbeck, Richard. 2012. "The Enduring Impact of the American Dust Bowl: Short- and Long-Run Adjustments to Environmental Catastrophe." *American Economic Review* 102, no. 4: 1477–507. www.nber.org/papers/w15605.pdf.

Horning, B. G., L. S. Morgan, B. H. Kneeland, and A. H. Hammar. 1941. "The Community Health Education Program: The Hartford Plan." *American Journal of Public Health* 31, no. 4: 310–18. https://ajph.aphapublications.org/doi/pdf/10.2105/ AJPH.31.4.310.

Howson, Henry. "American Jute." Philadelphia: The Franklin Institute, 1862. http:// galenet.galegroup.com/servlet/Sabin?af=RN&ae=CY100171476&srchtp=a&ste=14

Hu, Shiu-ying. 1955. *Flora of China, Family 153: Malvaceae*. Boston: Arnold Arboretum. http://hdl.handle.net/2027/coo.31924001223308.

Hughes, Harold De Mott, Maurice E. Heath, and Darrel S. Metcalfe. 1951. *Forages, the Science of Grassland Agriculture*. Ames: Iowa State College Press.

Hurt, R. Douglas. 1981. *The Dust Bowl: An Agricultural and Social History*. Chicago: Taylor.

Hutcheson, B. 2016. "Tiptonville Man Charged with Murder, Pleads Not Guilty." Newspaper. *Arkansas State Gazette* (blog). October 4. https://www.stategazette.com/ story/2344760.htm.

Hwang, Ji Hong, Qinglong Jin, Eun-Rhan Woo, and Dong Gun Lee. 2013. "Antifungal Property of Hibicuslide C and Its Membrane-Active Mechanism in Candida Albicans." *Biochimie* 95, no. 10: 1917–22. https://doi.org/10.1016/j.biochi.2013.06.019.

Hymowitz, T., and J. R. Harlan. 1983. "Introduction of Soybean to North America by Samuel Bowen in 1765." *Economic Botany* 37, no. 4: 371–79.

Iaffaldano, Brian. 2016. "Evaluating the Development and Potential Ecological Impact of Genetically Engineered Taraxacum Kok-Saghyz." The Ohio State University.

Integrated Taxonomic Information System (ITIS). 2019. "ITIS Standard Report Page: Setaria Faberi." Online database. https://www.itis.gov/.

Ioannou, D., J. Fortun, and H. G. Tempest. 2019. "Meiotic Nondisjunction and Sperm Aneuploidy in Humans." *Reproduction* 157: R15–31. https://doi.org/10.1530/ REP-18-0318.

Jasieniuk, M. 1995. "Constraints on the Evolution of Glyphosate Resistance in Weeds." *Resistant Pest Management Newsletter* 7, no. 2: 25–26.

Jefferson, Thomas. 1944. *Thomas Jefferson's Garden Book, 1766–1824, with Relevant Extracts from His Other Writings*. Philadelphia: American Philosophical Society.

Jefferson, Thomas. 1999. *Notes on the State of Virginia, 1785*. London: Penguin.

Jenkins, Virginia. 2015. *The Lawn: A History of an American Obsession*. Washington, DC: Smithsonian Institution.

Jensen, Merrill. 1969. "The American Revolution and American Agriculture." *Agricultural History* 43, no. 1: 107–24. http://www.jstor.org/stable/4617633.

Jing, Siqun, Saisai wang, Qian Li, Lian Zheng, Li Yue, Shaoli Fan, and Guanjun Tao. 2016. "Dynamic High Pressure Microfluidization-Assisted Extraction and Bioactivities of Cyperus Esculentus (C. Esculentus L.) Leaves Flavonoids." *Food Chemistry* 192: 319–27. https://doi.org/10.1016/j.foodchem.2015.06.097.

JinQing, Li, Su ZhiPing, Chen YanPing, Qi WeiZhen, Qi JiaLin, Duan XiaoHui, He LiNa, and Lu Min. 2017. "Analysis on Pests Intercepted Situation in Imported Renewable Resources at Shandong Port during 2006–2016." *Journal of Food Safety and Quality* 8, no. 11: 4120–24. https://www.cabdirect.org/cabdirect/abstract/20183006511.

Johnson, F. R. 1964. *The Peanut Story*. Murfreesboro, NC: Johnson.

Jomo, K. S., Anis Chowdhury, Krishnan Sharma, and Daniel Platz. 2016. "Public-Private Partnerships and the 2030 Agenda for Sustainable Development: Fit for Purpose?" New York: United Nations Department of Economic and Social Affairs. https://pdfs.semanticscholar.org/973a/1fe95a1b0f303eb5eaa50f6af07ae152c71b.pdf.

Jones, B. W. 1885. *The Peanut Plant: Its Cultivation and Uses . . .* New York: Orange Judd Company.

Jones, G. T., J. K. Norsworthy, and T. Barber. 2019. "Response of Soybean Offspring to a Dicamba Drift Event the Previous Year." *Weed Technology* 33, no. 1: 41–50. https://doi.org/doi: 10.1017/wet.2019.2.

Jordan, Carl F. 2002. "Genetic Engineering, the Farm Crisis, and World Hunger." *BioScience* 52, no. 6: 523. https://doi.org/10.1641/0006-3568(2002)052[0523:GETFCA]2.0.CO;2.

Kalm, Pehr. 1937. *The America of 1750: Peter Kalm's Travels in North America; the English Version of 1770, Rev. from the Original Swedish and Edited by Adolph B. Benson, with a Translation of New Material from Kalm's Diary Notes*. New York: Wilson-Erickson.

Kaufmann, Eric. 1999. "American Exceptionalism Reconsidered: Anglo-Saxon Ethnogenesis in the 'Universal' Nation, 1776–1850." *Journal of American Studies* 33, no. 3: 437–57. https://www.jstor.org/stable/27556685.

Kilpatrick, Jane. 2014. *Fathers of Botany: The Discovery of Chinese Plants by European Missionaries*. Chicago: University of Chicago Press.

Kiss, L., and Z. T. Beres. 2006. "Anthropogenic Factors behind the Recent Population Expansion of Common Ragweed (Ambrosia Artemisiifolia L.) in Eastern Europe: Is There a Correlation with Political Transitions?" *Journal of Biogeography* 33, no. 12: 2156–57. https://doi.org/10.1111/j.1365-2699.2006.01633.x.

Kline, David. 1990. *Great Possessions: An Amish Farmer's Journal*. San Francisco: North Point Press.

Kniss, Andrew R. 2017. "Long-Term Trends in the Intensity and Relative Toxicity of Herbicide Use." *Nature Communications* 8: 14865. https://doi.org/10.1038/ncomms14865.

Koo, Dal-Hoe, William T. Molin, Christopher A. Saski, Jiming Jiang, Karthik Putta, Mithila Jugulam, Bernd Friebe, and Bikram S. Gill. 2018. "Extrachromosomal Circular DNA-Based Amplification and Transmission of Herbicide Resistance in Crop Weed *Amaranthus Palmeri*." *Proceedings of the National Academy of Sciences* 115, no. 13: 3332–37. https://doi.org/10.1073/pnas.1719354115.

Koon, David. 2017. "Farmer vs. Farmer: The Fight over the Herbicide Dicamba Has Cost One Man His Life and Turned Neighbor against Neighbor in East Arkansas." Newspaper. *Arkansas Times* (blog). August 10. https://www.arktimes.com/arkansas/farmer-vs-farmer/Content?oid=8526754.

Kremer, Gary R. 2011. *George Washington Carver: A Biography*. Santa Barbara, CA: ABC-CLIO.

Kremer, Robert J. and Neal R. Spencer. 1989. "Impact of a Seed-Feeding Insect and Microorganisms on Velvetleaf (*Abutilon theophrasti*) Seed Viability." *Weed Science* 37: 211–16. http://www.jstor.org/stable/4044846.

Kroma, Margaret M, and Cornelia Butler Flora. 2003. "Greening Pesticides: A Historical Analysis of the Social Construction of Farm Chemical Advertisements." *Agriculture and Human Values* 20: 21–35.

Krstić, B., D. Stanković, R. Igić, and N. Nikolic. 2018. "The Potential of Different Plant Species for Nickel Accumulation." *Biotechnology and Biotechnological Equipment* 21, no. 4: 431–36. https://doi.org/10.1080/13102818.2007.10817489.

Kupperman, Karen Ordahl. 2012. *The Atlantic in World History*. New York: Oxford University Press.

Lapham, J., and D. S. H. Drennan. 1990. "The Fate of Yellow Nutsedge (Cyperus Esculentus) Seed and Seedlings in Soil." *Weed Science* 38: 125–28. https://doi.org/10.1017/S0043174500056253.

Latham, John. 1835. "Species of Trefoil or Clover." *Southern Agriculturist*, sec. 8.

Leiblein-Wild, Marion Carmen, Rana Kaviani, and Oliver Tackenberg. 2014. "Germination and Seedling Frost Tolerance Differ between the Native and Invasive Range in Common Ragweed." *Oecologia* 174, no. 3: 739–50. https://doi.org/10.1007/s00442-013-2813-6.

Liebman, M., C. L. Mohler, and C. P. Staver. 2001. *Ecological Management of Agricultural Weeds*. Cambridge: Cambridge University Press.

Lim, Dong-Ok, Ha-Song Kim, and Moon-Soo Park. 2009. "Distribution and Management of Naturalized Plants in the Northern Area of South Jeolla Province, Korea." *Korean Journal of Environment and Ecology* 23, no. 6: 506–15. http://www.koreascience.or.kr/article/JAKO200912368301273.page.

Linnaeus, C. von. 1753. *Species Plantarum*. 1st ed. Vol. 2. Stockholm: Laurentii Salvii.

Loomis, Dana, Kathryn Guyton, Yann Grosse, Fatiha El Ghissassi, Véronique Bouvard, Lamia Benbrahim-Tallaa, Neela Guha, Heidi Mattock, Kurt Straif, and International Agency for Research on Cancer Monograph Working Group, IARC, Lyon, France. 2015. "Carcinogenicity of Lindane, DDT, and 2,4-Dichlorophenoxyacetic Acid." *The Lancet. Oncology* 16, no. 8 (August): 891–92. https://doi.org/10.1016/S1470-2045(15)00081-9.

Lorenzi, H. 2000. "Plantas Daninhas Do Brasil: Terrestres, Parasitas, Aquáticas e Tóxicas." *Nova Odessa: Plantarum*: 90.

Loux, Mark M., and Mary Ann Berry. 1991. "Use of a Grower Survey for Estimating Weed Problems." *Weed Technology* 5, no. 2: 460–66. http://www.jstor.org/stable/3987462.

Madison, James. 1810. "From James Madison to Congress, 5 December 1810." Founders Online, National Archives. http://founders.archives.gov/documents/Madison/03-02-02-0192.

Madison, James. 1815. "Special Message to Congress February 20, 1815." Founders Online, National Archives. http://founders.archives.gov/documents/Madison/03-08-02-0206.

Makra, László, Miklós Juhász, Rita Béczi, and Emöke Borsos. 2005. "The History and Impacts of Airborne Ambrosia (Asteraceae) Pollen in Hungary." *Grana* 44, no. 1: 57–64. https://doi.org/10.1080/00173130510010558.

Makra, László, I. Matyasovszky, L. Hufnagel, and G. Tusdady. 2015. "The History of Ragweed in the World." *Applied Ecology and Environmental Research* 13, no. 2: 489–512. https://doi.org/10.15666/aeer/1302_489512.

Marcato, Ana Claudia de Castro, Cleiton Pereira de Souza, and Carmem Silvia Fontanetti. 2017. "Herbicide 2,4-D: A Review of Toxicity on Non-Target Organisms." *Water, Air, and Soil Pollution* 228, no. 3: 120. https://doi.org/10.1007/s11270-017-3301-0.

Marguiles, J. 2012. "No-Till Agriculture in the USA." In *Organic Fertilisation, Soil Quality, and Human Health*, edited by Eric Lichtfouse, 11–30. Dordrecht: Springer.

Markus, Catarine, Ales Pecinka, Ratna Karan, Jacob N. Barney, and Aldo Merotto. 2018. "Epigenetic Regulation—Contribution to Herbicide Resistance in Weeds? Epigenetic Regulation of Herbicide Resistance." *Pest Management Science* 74, no. 2: 275–81. https://doi.org/10.1002/ps.4727.

Mart, Michelle. 2015. *Pesticides, A Love Story: America's Enduring Embrace of Dangerous Chemicals.* Lawrence: University Press of Kansas. https://muse.jhu.edu/book/42335.

Martin, Alexander Campbell, Herbert Spencer Zim, and Arnold L. Nelson. 1961. *American Wildlife and Plants: A Guide to Wildlife Food Habits : The Use of Trees, Shrubs, Weeds, and Herbs by Birds and Mammals of the United States.* New York: Dover Publications.

Martin, Michael D., Elva Quiroz-Claros, Grace S. Brush, and Elizabeth A. Zimmer. 2018. "Herbarium Collection-Based Phylogenetics of the Ragweeds (Ambrosia, Asteraceae)." *Molecular Phylogenetics and Evolution* 120: 335–41. https://doi.org/10.1016/j.ympev.2017.12.023.

Mayer, Amy. 2018. "China Wants Fewer Weed Seeds in U.S. Soybeans It Imports." Iowa Public Radio. January. https://www.iowapublicradio.org/post/china-wants-fewer-weed-seeds-us-soybeans-it-imports.

Mears, S.M., and E. Charles. 1898. "More Fiber Plants: New Experiments with Oregon's Fertile Soil." *Oregonian.* March 14.

Medović, Aleksandar, and Ferenc Horváth. 2012. "Content of a Storage Jar from the Late Neolithic Site of Hódmezővásárhely-Gorzsa, South Hungary: A Thousand Carbonized Seeds of Abutilon Theophrasti Medic." *Vegetation History and Archaeobotany* 21, no. 3: 215–20. https://doi.org/10.1007/s00334-011-0319-x.

Miller, Philip. 1754. *The Gardeners Dictionary.* 4th ed. London: Printed for the author and sold by John and James Rivington. https://www.biodiversitylibrary.org/item/150892.

Mitich, Larry W. 1989. "Common Dandelion—the Lion's Tooth." *Weed Technology* 3, no. 3: 537–39. https://doi.org/10.1017/S0890037X00032735.

Mitman, Gregg. 2003. "Hay Fever Holiday: Health, Leisure, and Place in Gilded-Age America." *Bulletin of the History of Medicine* 77, no. 3: 600–635. https://www.jstor.org/stable/44447795.

Montagnani, C., R. Gentili, M. Smith, M.F. Guarino, and S. Citterio. 2017. "The Worldwide Spread, Success, and Impact of Ragweed (Ambrosia Spp.)." *Critical Reviews in Plant Sciences* 36, no. 3: 139–78. https://doi.org/10.1080/07352689.2017. 1360112.

Moreman, D. 2003. "Native American Ethnobotany Database." Native American Ethnobotany. http://naeb.brit.org/uses/search/?string=ambrosia.

Moretzsohn, Marcio de Carvalho, Mark S. Hopkins, Sharon E. Mitchell, Stephen Kresovich, Jose Francisco Montenegro Valls, and Marcio Elias Ferreira. 2004. "Genetic Diversity of Peanut (Arachis HypogaeaL.) and Its Wild Relatives Based on the Analysis of Hypervariable Regions of the Genome." *BMC Plant Biology* 4 (July 14): 11. https://doi.org/10.1186/1471-2229-4-11.

Morison, Samuel Eliot. 1950. *The Ropemakers of Plymouth: A History of the Plymouth Cordage Company, 1824–1949*. Boston: Houghton Mifflin.

Morison, Samuel Eliot. 1961. *The Maritime History of Massachusetts 1783–1860*. Boston: Houghton Mifflin.

Muhlenberg, Henry. 1793. "Index Florae Lancastriensis." *Transactions of the American Philosophical Society* 3: 157–84. https://doi.org/10.2307/1004866.

Murphy, Kathleen S. 2011. "Translating the Vernacular: Indigenous and African Knowledge in the Eighteenth-Century British Atlantic." *Atlantic Studies* 8, no. 1 (March 1): 29–48. https://doi.org/10.1080/14788810.2011.541188.

Nachtigal, Nicole C. 2001. "A Modern David and Goliath Farmer v. Monsanto: Advising a Grower on the Monsanto Technology Agreement." *Great Plains Natural Resources Journal* 6: 50. https://heinonline.org/HOL/Page?handle=hein.journals/gpnat6&id=56&div=&collection=.

Nandula, Vijay K., Thomas W. Eubank, Daniel H. Poston, Clifford H. Koger, and Krishna N. Reddy. 2006. "Factors Affecting Germination of Horseweed (Conyza Canadensis)." *Weed Science* 54, no. 5: 898–902. https://doi.org/10.1614/WS-06-006R2.1.

National Academies of Sciences, Engineering, and Medicine. 2016. "Committee on Gene Drive Research in Non-Human Organisms: Recommendations for Responsible Conduct; Board on Life Sciences; Division on Earth and Life Studies; National Academies of Sciences, Engineering, and Medicine. Gene Drives on the Horizon: Advancing Science, Navigating Uncertainty, and Aligning Research with Public Values." Washington, DC: National Academies of Sciences, Engineering, and Medicine. https://www.ncbi.nlm.nih.gov/books/NBK379273/.

Negbi, M. 1992. "A Sweetmeat Plant, a Perfume Plant and Their Weedy Relatives: A Chapter in the History of Cyperus Esculentus L and C. Rotundus L." *Economic Botany* 46, no. 1: 64–71. https://doi-org.proxy.lib.ohio-state.edu/10.1007/BF02985255.

Nesom, G.L. 1990. "Further Definition of Conyza (Asteraceae: Astereae)." *Phytologia* 68, no. 3: 229–33. https://doi.org/10.5962/bhl.part.16712.

Neve, Paul. 2018. "Gene Drive Systems: Do They Have a Place in Agricultural Weed Management?" *Pest Management Science* 74, no. 12: 2671–79. https://doi.org/10.1002/ps.5137.

New Jersey, State of. 1881. "Annual Report New Jersey Bureau of Statistics, Labor, and Industries." Annual. Trenton: State of New Jersey. https://books.google.com/books?id=lwZbAAAAYAAJ.

Noyes, Richard D. 2007. "Apomixis in the Asteraceae: Diamonds in the Rough." *Functional Plant Science and Biotechnology* 1, no. 2: 207–22.

Nurse, R. E., S. J. Darbyshire, C. Bertin, and A. DiTommaso. 2009. "The Biology of Canadian Weeds. 141. Setaria Faberi Herrm." *Canadian Journal of Plant Science* 89, no. 2: 379–404. https://doi.org/10.4141/CJPS08042.

O'Connor, C. 2008. "Meiosis, Genetic Recombination, and Sexual Reproduction." *Nature Education* 1, no. 1: 174.

Oh, Choong-Hyeon, Yong-Hoon Kim, Lee Ho-Young, and Su-Hong Ban. 2009. "The Naturalization Index of Plant around Abandoned Military Camps in Civilian Control Zone." *Journal of the Korean Society of Environmental Restoration Technology* 12, no. 5: 59–76. http://ocean.kisti.re.kr/downfile/volume/kserrt/HKBOB5/2009/v12n5/HKBOB5_2009_v12n5_59.pdf.

Okoli, C.A.N., D. G. Shilling, R. L. Smith, and T. A. Bewick. 1997. "Genetic Diversity in Purple Nutsedge (Cyperus RotundusL.) and Yellow Nutsedge (Cyperus EsculentusL.)." *Biological Control* 8, no. 2: 111–18. https://doi.org/10.1006/bcon.1996.0490.

Olabiyi, Ayodeji Augustine, Ganiyu Oboh, and Stephen Adeniyi Adefegha. 2017. "Effect of Dietary Supplementation of Tiger Nut (Cyperus Esculentus l.) and Walnut (Tetracarpidium Conophorum Müll. Arg.) on Sexual Behavior, Hormonal Level, and Antioxidant Status in Male Rats." *Journal of Food Biochemistry* 41, no. 3: e12351. https://doi.org/10.1111/jfbc.12351.

Ortega, Suzanne T., David R. Johnson, Peter G. Beeson, and Betty J. Craft. 1994. "The Farm Crisis and Mental Health: A Longitudinal Study of the 1980s." *Rural Sociology* 59, no. 4: 598–619. https://doi.org/10.1111/j.1549-0831.1994.tb00550.x.

Osband, Kent. 1985. "The Boll Weevil Versus 'King Cotton.'" *Journal of Economic History* 45, no. 3: 627–43. http://www.jstor.org.proxy.lib.ohio-state.edu/stable/2121755.

Padgette, Stephen R., K. H. Kolacz, X. Delannay, D.B. Re, B.J. LaVallee, C.N. Tinius, W.K. Rhodes, Y.I. Otero, G.F. Barry, and D.A. Eichholtz. 1995. "Development, Identification, and Characterization of a Glyphosate-Tolerant Soybean Line." *Crop Science* 35, no. 5: 1451–61.

Palhano, M.G., J. K. Norsworthy, and T. Barber. 2018. "Cover Crops Suppression of Palmer Amaranth (Amaranthus Palmeri) in Cotton." *Weed Technology* 32, no. 1: 60–65.

Panero, Jose L., Susana E. Freire, Luis Ariza Espinar, Bonnie S. Crozier, Gloria E. Barboza, and Juan J. Cantero. 2014. "Resolution of Deep Nodes Yields an Improved Backbone Phylogeny and a New Basal Lineage to Study Early Evolution of Asteraceae." *Molecular Phylogenetics and Evolution* 80: 43–53. https://doi.org/10.1016/j.ympev.2014.07.012.

Pascual, B., J.V. Maroto, S. Lopez-Galarza, A. Sanbautista, and J. Alagarda. 2000. "Chufa (Cyperus Esculentus L. Var. Satifus Boeck.): An Unconventional Crop. Studies Related to Applications and Cultivation." *Economic Botany* 54, no. 4: 439–48. https://doi-org.proxy.lib.ohio-state.edu/10.1007/BF02866543.

Pasqualini, Stefania, Emma Tedeschini, Giuseppe Frenguelli, Nicole Wopfner, Fatima Ferreira, Gennaro D'Amato, and Luisa Ederli. 2011. "Ozone Affects Pollen Viability and NAD(P)H Oxidase Release from Ambrosia Artemisiifolia Pollen." *Environmental Pollution* 159, no. 10: 2823–30. https://doi.org/10.1016/j.envpol.2011.05.003.

Peacock, L.N. 2018. "EPA Ignores Its Scientists on Dicamba Buffers; New Rules May Allow Dicamba Use in Arkansas." *Arkansas Times* (blog). November 21. https://www.arktimes.com/ArkansasBlog/archives/2018/11/21/epa-ignores-its-scientists-on-dicamba-buffers-new-rules-may-allow-dicamba-use-in-arkansas.

Perry, Melissa J. 2008. "Effects of Environmental and Occupational Pesticide Exposure on Human Sperm: A Systematic Review." *Human Reproduction Update* 14, no. 3 (May 1): 233–42. https://doi.org/10.1093/humupd/dmm039.

Peterson, D., M. Jugulam, C. Shyam, and E. Borgato. 2019. "Palmer Amaranth Resistance to 2,4-D and Dicamba Confirmed in Kansas." Kansas State University. K-State Agronomy eUpdates, March 1. https://webapp.agron.ksu.edu/agr_social/m_eu_article.throck?article_id=2110&eu_id=322.

Peterson, Gale. E. 1967. "The Discovery and Development of 2,4-D." *Agricultural History* 41, no. 3: 243–54. https://www-jstor-org.proxy.lib.ohio-state.edu/stable/3740338.

Pickering, Charles. 1879. *Chronological History of Plants: Man's Record of His Own Existence Illustrated through Their Names, Uses, and Companionship.* Boston: Little, Brown.

Pohl, R.W. 1951. "The Genus Setaria in Iowa." *Iowa State College Journal of Science.* 25: 501–8.

Price, A.J., C.D. Monks, A.S. Culpepper, L.M. Duzy, J.A. Kelton, M.W. Marshall, L.E. Steckel, L.M. Sosnoskie, and R.L. Nichols. 2016. "High-Residue Cover Crops Alone or with Strategic Tillage to Manage Glyphosate-Resistant Palmer Amaranth (Amaranthus Palmeri) in Southeastern Cotton (Gossypium Hirsutum)." *Journal of Soil and Water Conservation* 71, no. 1: 1–11. https://doi.org/10.2489/jswc.71.1.1.

Rasmussen, Nicolas. 2001. "Plant Hormones in War and Peace: Science, Industry, and Government in the Development of Herbicides in 1940s America." *Isis* 92, no. 2 (June 1): 291–316. https://doi.org/10.1086/385183.

Rastogi, S.M., M. Pandey, and A.K.S. Rawat. 2011. "An Ethnomedicinal, Phytochemical and Pharmacological Profile of Desmodium Gangeticum (L.) DC. and Desmodium Adscendens (Sw.) DC." *Journal of Ethnopharmacology* 136: 283–96.

Ratner, Sidney. 1972. *The Tariff in American History.* New York: Van Nostrand.

Regnier, E., S.K. Harrison, J. Liu, J.T. Schmoll, C.A. Edwards, N. Arancon, and C. Holloman. 2008. "Impact of an Exotic Earthworm on Seed Dispersal of an Indigenous US Weed." *Journal of Applied Ecology* 45, no. 6: 1621–29. https://doi.org/10.1111/j.1365-2664.2008.01489.x.

Regnier, Emilie E., S. Kent Harrison, Mark M. Loux, Christopher Holloman, Ramarao Venkatesh, Florian Diekmann, Robin Taylor et al. 2016. "Certified Crop Advisors' Perceptions of Giant Ragweed (Ambrosia Trifida) Distribution, Herbicide Resistance, and Management in the Corn Belt." *Weed Science* 64, no. 2: 361–77. https://doi.org/10.1614/WS-D-15-00116.1.

Reznik, Sergey Ya. 2009. "Common Ragweed (Ambrosia Artemisiifolia L.) in Russia: Spread, Distribution, Abundance, Harmfulness and Control Measures." *Ambrosie* 26: 88–97. http://www.zin.ru/labs/expent/pdfs/Reznik_2009_Ambrosia.pdf.

Ribeiro, Daniela N., Zhiqiang Pan, Stephen O. Duke, Vijay K. Nandula, Brian S. Baldwin, David R. Shaw, and Franck E. Dayan. 2014. "Involvement of Facultative Apomixis in Inheritance of EPSPS Gene Amplification in Glyphosate-Resistant Amaranthus Palmeri." *Planta* 239, no. 1: 199–212. https://doi.org/10.1007/s00425-013-1972-3.

Richards, A.J. 1973. "The Origin of Taraxacum Agamospecies." *Botanical Journal of the Linnean Society* 66, no. 3: 189–211. https://doi.org/10.1111/j.1095-8339.1973.tb02169.x.

Robbins, Paul. 2012. *Lawn People: How Grasses, Weeds, and Chemicals Make Us Who We Are.* Philadelphia: Temple University Press.

Robbins, Wilfred W., Alden S. Crafts, and Richard N. Raynor. 1952. *Weed Control*, 2nd ed. New York: McGraw-Hill.

Rodney, Walter. 1970. *A History of the Upper Guinea Coast, 1545–1800.* Oxford Studies in African Affairs. Oxford: Clarendon.

Rogers, Christine A., Peter M. Wayne, Eric A. Macklin, Michael L. Muilenberg, Christopher J. Wagner, Paul R. Epstein, and Fakhri A. Bazzaz. 2006. "Interaction of the Onset of Spring and Elevated Atmospheric CO2 on Ragweed (Ambrosia Artemisiifolia L.) Pollen Production." *Environmental Health Perspectives* 114, no. 6: 865–69. https://doi.org/10.1289/ehp.8549.

Rogin, Michael Paul. 1991. *Fathers and Children: Andrew Jackson and the Subjugation of the American Indian.* New Brunswick, NJ: Transaction Publishers.

Romans, Bernard. (1775) 1962. *A Concise Natural History of East and West Florida.* Floridiana Facsimile and Reprint Series. Gainesville: University of Florida Press. https://catalog.hathitrust.org/Record/001264063.

Rominger, James M. 1962. "Taxonomy of Setaria (Gramineae) in North America." *Illinois Biological Monographs* 29: 156. https://archive.org/details/taxonomyofsetari29romi.

Rüegg, W. T., M. Quadranti, and A. Zoschke. 2007. "Herbicide Research and Development: Challenges and Opportunities." *Weed Research* 47, no. 4: 271–75. https://doi.org/10.1111/j.1365-3180.2007.00572.x.

Rutherford, P.P., and A.C. Deacon. 1974. "Seasonal Variation in Dandelion Roots of Fructosan Composition, Metabolism, and Response to Treatment with 2,4-Dichlorophenoxyacetic Acid." *Annals of Botany* 38, no. 2: 251–60. https://doi.org/10.1093/oxfordjournals.aob.a084809.

S.W.B. 1878. "Editorial Correspondence." *Georgia Weekly Telegraph.* November 19.

Sammons, Robert Douglas, and Todd A. Gaines. 2014. "Glyphosate Resistance: State of Knowledge." *Pest Management Science* 70, no. 9: 1367–77. https://doi.org/10.1002/ps.3743.

Sato, Yukie, Yuta Mashimo, Ryo O. Suzuki, Akira S. Hirao, Etsuro Takagi, Ryuji Kanai, Daisuke Masaki, Miyuki Sato, and Ryuichiro Machida. 2017. "Potential Impact of an Exotic Plant Invasion on Both Plant and Arthropod Communities in a Semi-Natural Grassland on Sugadaira Montane in Japan." *Journal of Developments in Sustainable Agriculture* 12: 1352–64. https://doi-org.proxy.lib.ohio-state.edu/10.11178/jdsa.12.52.

Sauer, J. D. 1957. "Recent Migration and Evolution of the Dioecious Amaranths." *Evolution* 11, no. 1: 11–31. https://doi.org/10.1111/j.1558-5646.1957.tb02872.x.

Sauer, J. D. 1967. "The Grain Amaranths and Their Relatives: A Revised Taxonomic and Geographic Survey." *Annals of the Missouri Botanical Garden* 54, no. 2: 102–37. https://www.jstor.org/stable/2394998.

Sauer, J. D. 1988. *Plant Migration: The Dynamics of Geographic Patterning in Seed Plant Species*. Berkeley: University of California Press.

Scheepens, P. C., and A. Hoogerbrugge. 1991. "Host Specificity of Puccinia Canaliculata, a Potential Biocontrol Agent for Cyperus Esculentus." *Netherlands Journal of Plant Pathology* 97, no. 4: 245–50. https://doi.org/10.1007/BF01989821.

Schippers, Peter, Siny J. Ter Borg, and Jan Just Bos. 1995. "A Revision of the Infraspecific Taxonomy of Cyperus Esculentus (Yellow Nutsedge) with an Experimentally Evaluated Character Set." *Systematic Botany* 20, no. 4: 461–81. https://www.jstor.org/stable/2419804.

Schubert, B. G. 1945. "Flora of Peru: Desmodium Desv." *Publications of Field Museum of Natural History* 13: 413–39.

Selby, A. D. 1897. *A First Ohio Weed Manual*. Bulletin of the Ohio Agricultural Experiment Station 83. Wooster, OH: Ohio Agricultural Experiment Station.

Sharma, Budh Dev. 1993. *Flora of India. 3. Portulacaceae-Ixonanthaceae*. Totnes, Devon, UK.: Botanical survey of India.

Sheahan, Megan, Christopher B. Barrett, and Casey Goldvale. 2017. "Human Health and Pesticide Use in Sub-Saharan Africa." *Agricultural Economics* 48, no. S1: 27–41. https://doi.org/10.1111/agec.12384.

Simberloff, Daniel. 2006. "Invasional Meltdown 6 Years Later: Important Phenomenon, Unfortunate Metaphor, or Both?" *Ecology Letters* 9: 912–19. https://doi.org/10.1111/j.1461-0248.2006.00939.x.

Simpson, C. E., A. Krapovickas, and J. F. M. Valls. 2001. "History of Arachis Including Evidence of A. Hypogaea L. Progenitors." *Peanut Science* 28, no. 2 (July 1): 78–80. https://doi.org/10.3146/i0095-3679-28-2-7.

Sloane, Hans. 1707. *A Voyage to the Islands Madera, Barbados, Nieves, S. Christophers and Jamaica: With the Natural History . . . of the Last of Those Islands; to Which Is Prefix'd an Introduction, Wherein Is an Account of the Inhabitants, Air, Waters, Diseases, Trade, &c. . . . Illustrated with the Figures of the Things Describ'd, . . . By Hans Sloane, . . . In Two Volumes*. Vol. 2. London: printed by B. M. for the author.

Smith, Andrew F. 2002. *Peanuts: The Illustrious History of the Goober Pea*. The Food Series. Urbana: University of Illinois Press.

Smith, F. G., C. L. Hamner, and R. F. Carlson. 1946. "Control of Ragweed Pollen Production with 2,4-Dichlorophenoxyacetic Acid." *Science* 103, no. 2677: 473–74. https://doi.org/10.1126/science.103.2677.473.

Smith, Jared Gage. 1896. *Fodder and Forage Plants: Exclusive of the Grasses*. U.S. Department of Agriculture, Division of Agrostology.

Smith, P. 2016. "Dicamba: The 'Time Bomb' Went Off and No One Was Prepared." Magazine. *AGFACTS: Progressive Farmer* (blog), December 29. https://agfax.com/2016/12/29/dicamba-the-time-bomb-went-off-and-no-one-was-prepared-dtn/.

Solbrig, Otto T. 1971. "The Population Biology of Dandelions." *American Scientist* 59 \: 686–94. http://adsabs.harvard.edu/abs/1971AmSci..59..686S.

Sosnoskie, L. M., T. M. Webster, D. Dales, G. C. Rains, T. L. Grey, and A. S. Culpepper. 2009. "Pollen Grain Size, Density, and Settling Velocity for Palmer Amaranth (Amaranthus Palmeri)." *Weed Science* 57, no. 04: 404–9. https://doi.org/10.1614/WS-08-157.1.

Spence, R. A. 1902. "Social Rise of the Peanut." *Good Housekeeping.*

Steckel, Lawrence E. 2007. "The Dioecious Amaranthus Spp.: Here to Stay." *Weed Technology* 21, no. 2: 567–70. https://doi.org/10.1614/WT-06-045.1.

Steffen, Will, Johan Rockström, Katherine Richardson, Timothy M. Lenton, Carl Folke, Diana Liverman, Colin P. Summerhayes et al. 2018. "Trajectories of the Earth System in the Anthropocene." *Proceedings of the National Academy of Sciences* 115, no. 33: 8252–59. https://doi.org/10.1073/pnas.1810141115.

Stępalska, Danuta, Dorota Myszkowska, Leśkiewicz Katarzyna, Piotrowicz Katarzyna, Borycka Katarzyna, Chłopek Kazimiera, Grewling Łukasz et al. 2017. "Co-Occurrence of Artemisia and Ambrosia Pollen Seasons against the Background of the Synoptic Situations in Poland." *International Journal of Biometeorology* 61, no. 4: 747–60. https://doi.org/10.1007/s00484-016-1254-4.

Stepanek, L., J. Evertson, and K. Martens. 2018. "Herbicide Damage to Trees." University of Nebraska. Nebraska Forest Service, April 18. https://nfs.unl.edu/publications/herbicide-damage-trees.

Stewart-Wade, S. M., S. Neumann, L. L. Collins, and G. J. Boland. 2002. "The Biology of Canadian Weeds. 117. Taraxacum Officinale G. H. Weber Ex Wiggers." *Canadian Journal of Plant Science* 82, no. 4: 825–53. https://doi.org/10.4141/P01-010.

Stoller, E. W. 1981. "Yellow Nutsedge: A Menace in the Corn Belt." USDA Technical Bulletin. Washington, DC: U.S. Department of Agriculture, Economic Research Service.

Sturtevant, E. Lewis. 1886. "A Study of the Dandelion." *American Naturalist* 20, no. 1: 5–9.

Swartz, Olof Peter. 1788. *Nova genera and species plantarum seu prodromus descriptionum vegetalium, maximam partem incognitorum quæ sub itinere in Indiam occidentalem annis 1783–1787 digessit Olof Swartz. M.D.* in bibliopolis Acad. M. Swederi.

Swingle, Walter T. 1945. "Our Agricultural Debt to Asia." In *The Asian Legacy and American Life*, 3rd ed., edited by Arthur E. Christy, 84–114. Toronto: Asia Press.

Tabuti, J. R. S, K. A Lye, and S. S. Dhillion. 2003. "Traditional Herbal Drugs of Bulamogi, Uganda: Plants, Use, and Administration." *Journal of Ethnopharmacology* 88, no. 1 (September): 19–44. https://doi.org/10.1016/S0378-8741(03)00161-2.

Tandyekkal, Dhruvan. 1997. "Desmodium Tortuosum (Sw.) DC. (Fabaceae): A New Record for Kerala." *Journal of Economic and Taxonomic Botany* 21, no. 3: 663–66. https://www.cabdirect.org/cabdirect/abstract/19982303287.

Taramarcaz, P., C. Lambelet, B. Clot, C. Keimer, and C. Hauser. 2005. "Ragweed (Ambrosia) Progression and Its Health Risks: Will Switzerland Resist This Invasion?" *Swiss Medical Weekly* 135, no. 3738: 538–48. https://smw.ch/journalfile/view/article/ezm_smw/en/smw.2005.11201/de6402c6a28db5f0ccc2f94b28c2ae1fc9feddd1/smw.2005.11201.pdf/rsrc/jf.

Thoreau, Henry David. 1962. *Walden*. New York: Time Inc.

Thoreau, Henry David. 1993. *Faith in a Seed: The Dispersion of Seeds*. Washington, DC: Island Press.

Thoreau, Henry David. 2001. *Wild Fruits: Thoreau's Rediscovered Last Manuscript*. New York: W.W. Norton.

Thullen, R.J., and P.E. Keeley. 1979. "Seed Production and Germination in Cyperus Esculentus and C. Rotundus." *Weed Science* 27, no. 5: 502–5. https://doi.org/10.1017/S0043174500044489.

Timmons, F.L. 1970. "A History of Weed Control in the United States and Canada." *Weed Science* 18, no. 2: 294–307. https://doi.org/10.1017/S0043174500079807.

Troyer, James R. 2001. "In the Beginning: The Multiple Discovery of the First Hormone Herbicides." *Weed Science* 49, no. 2: 290–97. https://doi.org/10.1614/0043-1745 (2001)049[0290:ITBTMD]2.0.CO;2.

Trucco, Federico, and Patrick J. Tranel. 2011. "Amaranthus." In *Wild Crop Relatives: Genomic and Breeding Resources*, edited by Chittaranjan Kole, 11–21. Berlin: Springer. https://doi.org/10.1007/978-3-642-20450-0_2.

Tsai, Jen-Chieh. 2011. "Antioxidant Activities of Phenolic Components from Various Plants of Desmodium Species." *African Journal of Pharmacy and Pharmacology 5*, no. 4 (April 30): 468–76. https://doi.org/10.5897/AJPP11.059.

Tumbleson, M.E., and Thor Kommedahl. 1961. "Reproductive Potential of Cyperus Esculentus by Tubers." *Weeds 9*, no. 4: 646–53. https://doi.org/10.2307/4040817.

Unglesbee, Emily. 2019. "How Dicamba's Visibility Could Change Ag Pesticide Use Forever." Farmer magazine. *Production Blog* (blog), March 27. https://www.dtnpf.com/agriculture/web/ag/perspectives/blogs/production-blog/blog-post/2019/03/27/dicambas-visibility-change-ag-use.

United Nations. 2009. Department of Public Information. *Millennium Development Goals Report 2009*. New York: United Nations Publications.

U.S. Department of Agriculture. PLANTS database. https://plants.sc.egov.usda.gov/.

U.S. Department of Agriculture. 1942. "Soybean Oil and the War: Grow More Soybeans for Victory." USDA Bureau of Agricultural Economics, Extension Flier. No. 5.

Valencius, C.B. 2002. *The Health of the Country: How American Settlers Understood Themselves and Their Land*. New York: Basic Books.

Van Dijk, Peter. 2009. "Apomixis: Basics for Non-Botanists." In *Lost Sex*, edited by Isa Schön, Koen Martens, and Peter Van Dijk, 47–62. Dordrecht: Springer.

VanGessel, Mark J. 2001. "Glyphosate-Resistant Horseweed from Delaware." *Weed Science* 49, no. 6: 703–5. https://doi.org/www.jstor.org/stable/4046416.

Van Horn, Christopher R., Marcelo L. Moretti, Renae R. Robertson, Kabelo Segobye, Stephen C. Weller, Bryan G. Young, William G. Johnson et al. 2018. "Glyphosate Resistance in Ambrosia Trifida: Part 1. Novel Rapid Cell Death Response to Glyphosate." *Pest Management Science* 74, no. 5:1071–078. http://onlinelibrary.wiley.com/doi/abs/10.1002/ps.4567

Van Wychen, L. 2018. "Survey of the Most Common and Troublesome Weeds in Broadleaf Crops, Fruits and Vegetables in the United States and Canada." *Weed Science*

Society of America (blog). http://wssa.net/wp-content/uploads/2016-Weed-Survey_
Broadleaf-crops.xlsx.

Vavílov, N. I. 1992. *Origin and Geography of Cultivated Plants.* New York: Cambridge University Press.

Venables, Robert W. 2004. *American Indian History: Five Centuries of Conflict and Coexistence.* Santa Fe, NM: Clear Light Publishers.

Vries, Femke T. de. 1991. "Chufa (Cyperus Esculentus, Cyperaceae): A Weedy Cultivar or a Cultivated Weed?" *Economic Botany* 45, no. 1: 27–37. https://doi.org/10.1007/BF02860047.

Waite, Kathryn J. 1995. "Blackley and the Development of Hay Fever as a Disease of Civilization in the Nineteenth Century." *Medical History* 39, no. 2: 186–96. https://doi.org/10.1017/S0025727300059834.

Walzer, M., and B. B. Siegel. 1956. "The Effectiveness of the Ragweed Eradication Campaigns in New York City: A 9-Year Study (1946–1954)." *Journal of Allergy* 27, no. 2: 113–26. https://doi.org/10.1016/0021-8707(56)90002-8.

Wang, Rong-Lin, Jonathan F. Wendel, and Jack H. Dekker. 1995. "Weedy Adaptation in Setaria Spp. II. Genetic Diversity and Population Genetic Structure in S. Glauca, S. Geniculata, and S. Faberii (Poaceae)." *American Journal of Botany* 82, no. 8: 1031–39. https://doi.org/10.2307/2446233.

Ward, Sarah M., Theodore M. Webster, and Larry E. Steckel. 2013. "Palmer Amaranth (Amaranthus Palmeri): A Review." *Weed Technology* 27, no. 01: 12–27. https://doi.org/10.1614/WT-D-12-00113.1.

Watts, Meriel. 2011. "Paraquat." Penang, Malaysia: Pesticide Action Network Asia and the Pacific. http://wssroc.agron.ntu.edu.tw/note/Paraquat.pdf.

Wayne, Peter, Susannah Foster, John Connolly, Fakhri Bazzaz, and Paul Epstein. 2002. "Production of Allergenic Pollen by Ragweed (Ambrosia Artemisiifolia L.) Is Increased in CO_2-Enriched Atmospheres." *Annals of Allergy, Asthma and Immunology* 88, no. 3: 279–82. https://doi.org/10.1016/S1081-1206(10)62009-1.

Webster, Theodore M., and John Cardina. 2004. "A Review of the Biology and Ecology of Florida Beggarweed (Desmodium Tortuosum)." *Weed Science* 52, no. 2: 185–200. http://www.wssajournals.org/doi/abs/10.1614/WS-03-028R.

Webster, Theodore M., and Harold D. Coble. 1997. "Changes in the Weed Species Composition of the Southern United States: 1974 to 1995." *Weed Technology* 11, no. 2: 308–17. http://www.jstor.org/stable/3988731.

Webster, Theodore M., and Robert L. Nichols. 2012. "Changes in the Prevalence of Weed Species in the Major Agronomic Crops of the Southern United States: 1994/1995 to 2008/2009." *Weed Science* 60, no. 2: 145–57. http://www.jstor.org.proxy.lib.ohio-state.edu/stable/pdf/41497617.pdf.

Webster, Theodore M., and Lynn M. Sosnoskie. 2010. "Loss of Glyphosate Efficacy: A Changing Weed Spectrum in Georgia Cotton." *Weed Science* 58, no. 1: 73–79. https://doi.org/10.1614/WS-09-058.1.

Weed Science Society of America "Composite List of Weeds." http://wssa.net/wssa/weed/composite-list-of-weeds/.

Weiner, J., and R. P. Freckleton. 2010. "Constant Final Yield." *Annual Review of Ecology, Evolution, and Systematics* 41: 173–92. www.jstor.org/stable/27896219.

Whaley, W. G., and J. S. Bowen. 1947. *Russian Dandelion (Kok-Saghyz): An Emergency Source of Natural Rubber*. Washington, DC: U.S. Department of Agriculture.

Whitney, G. G. 1996. *From Coastal Wilderness to Fruited Plain: A History of Environmental Change*. Cambridge: Cambridge University Press.

Wilbur, Robert L. 1963. "The Leguminous Plants of North Carolina." *Technical Bulletin/North Carolina Agricultural Experiment Station*, Tech. bul./North Carolina Agricultural Experiment Station; no. 151, 151: 294.

Wilcut, John W., Glenn R. Wehtje, Tracy A. Cole, T. Vint Hicks, and John A. McGuire. 1989. "Postemergence Weed Control Systems without Dinoseb for Peanuts (Arachis Hypogaea)." *Weed Science* 37, no. 3: 385–91. http://www.jstor.org/stable/4044727.

Willingham, Samuel D., Barry J. Brecke, Joyce Treadaway-Ducar, and Gregory E. Mac-Donald. 2008. "Utility of Reduced Rates of Diclosulam, Flumioxazin, and Imazapic for Weed Management in Peanut." *Weed Technology* 22, no. 1: 74–80. https://doi.org/10.1614/WT-07-074.1.

Wright, S. I., S. Kalisz, and T. Slotte. 2013. "Evolutionary Consequences of Self-Fertilization in Plants." *Proceedings of the Royal Society B: Biological Sciences* 280, no. 1760 (April 17): 20130133-20130133. https://doi.org/10.1098/rspb.2013.0133.

Wulf, Andrea. 2015. *The Invention of Nature: Alexander Von Humboldt's New World*. London: John Murray.

Yakkala, K., M. R. Yu, J. K. Yang, and Y. Y. Chang. 2013. "Adsorption of TNT and RDX Contaminants by Ambrosia Trifida L. Var. Trifida Derived Biochar." *Research Journal of Chemistry and Environment* 17, no. 4: 62–71. https://www.scopus.com/inward/record.uri?eid=2-s2.0-84880684950&partnerID=40&md5=284ce3c00b700a735c4535b2d5545536.

Yuan, Dawei, Ludovic Bassie, Maite Sabalza, Bruna Miralpeix et al. 2011. "The Potential Impact of Plant Biotechnology on the Millennium Development Goals." *Plant Cell Reports* 30, no. 3: 249–65. https://doi.org/10.1007/s00299-010-0987-5.

Zeisberger, David, A. B. Hulbert, and W. N. Schwarze. 1910. *David Zeisberger's History of the Northern American Indians*. Columbus, OH: Ohio State Archaeological and Historical Society. http://hdl.handle.net/2027/uiug.30112003956056.

Zhang, Yingxiao, Brian J. Iaffaldano, Xiaofeng Zhuang, John Cardina, and Katrina Cornish. 2017. "Chloroplast Genome Resources and Molecular Markers Differentiate Rubber Dandelion Species from Weedy Relatives." *BMC Plant Biology* 17, no. 1: 34. https://doi.org/10.1186/s12870-016-0967-1.

Zimdahl, R. L. 1989. *Weeds and Words. The Etymology of the Scientific Names of Weeds and Crops*. Ames: Iowa State University Press.

Ziska, Lewis H., Dennis E. Gebhard, David A. Frenz, Shaun Faulkner, Benjamin D. Singer, and James G. Straka. 2003. "Cities as Harbingers of Climate Change: Common Ragweed, Urbanization, and Public Health." *Journal of Allergy and Clinical Immunology* 111, no. 2: 290–95. https://doi.org/10.1067/mai.2003.53.

Ziska, Lewis H., Kim Knowlton, Christine A. Rogers, Dan Dalan, Nicole Tierney, Mary Ann Elder, Warren Filley et al. 2011. "Recent Warming by Latitude Associated with

Increased Length of Ragweed Pollen Season in Central North America." *Proceedings of the National Academy of Sciences* 108, no. 10: 4248–51. https://doi-org.proxy.lib. ohio-state.edu/10.1073/pnas.1014107108.

Zohary, D., and M. Hopf. 2000. *Domestication of Plants in the Old World: The Origin and Spread of Cultivated Plants in West Asia, Europe, and the Nile Valley.* Vol. 2. Oxford: Oxford University Press.

INDEX

Note: *Italicized* page numbers indicate illustrations

Abu Ali Sina, 76
Abuliton genus, 75
Abutilon avicennae, 76, 85, 88, 90
"Abutilon hemp," 76, 78, 80
Agrestal selection
  agriculture and human interventions, 2, 9–11, 13, 14, 238
  giant foxtail and, 227, 231, 232
  giant ragweed and, 199, 204
  marestail and, 140, 147–48, 151
  pigweed and, 161, 163–65, 166, 174, 180, 182–83
  velvetleaf and, 84, 92, 98–100
  yellow nutsedge and, 109, 114
Agriculture, generally
  agrestal selection, 2, 9–11, 13, 14, 238
  "climate-smart" agriculture, 238
  food policy and the future, 234–39

slavery and weed removal, 8–9
weeds' role in settled agriculture, 6–7, 11–13
"Agro-dealers," in Tanzania, 110–12, 121, 123, 125
Alachlor herbicide, 93, 96, 135, 213
Alexander Seed Company, 62
Allergies. *See* Pollen, giant ragweed and allergenicity
ALS herbicides, pigweed and, 163–65
Alternative farmers, 14, 136–39, 152–54
American jute. *See* Velvetleaf *(Abutilon theophrasti)*
Amish farms, management of, 14, 225–27
Anthropocene, and weeds generally, 8–9, 12–13, 207–9, 238
Anti-GMO movement, 154–55

Apomixis
    dandelions and, 18–21, 25–26, 30,
        33, 42
    pigweed and, 181–82
Asteraceae family, 18–19, 187
Atrazine herbicide, 92–93, 96, 135
Auxin herbicides, 36. *See also*
    2,4-dichlorophenoxyacetic acid
    (2,4-D)

Banister, John Baptist, 77
Barton, Benjamin Smith, 78
Bartram, John, 160, 193
Bartram, William, 57–58
Baumann, Hermann, 116
Beard, George M., 195
Bernfeld, Siegfried, 32
Biotechnology, genetics of plants and,
    140–42
Blackley, Charles, 194–95
Bostock, John, 194, 195
Bradley, Kevin, 179
Brown, Samuel C., 86–87, 88

C3 plants, 207
C4 plants, 214
Candolle, Alphonse de, 56
Cannabis sativa ("true" hemp), 4
Carver, Dr. George Washington, 63
China
    dandelions and, 41–42
    foxtail and, 214, 215–16
    ragweed and, 201, 205–6
    velvetleaf and, 75–77, 91, 99
Chufa ("tiger nut"), 108–10, 117–18, 120
Cis-1,4-polyisoprene, 41–43
Civil War, American, 25, 60, 86
"Classic" pigweeds, 159–60, 161.
    See also Pigweed
Clayton, John, 77
Climate change, 1, 13
    dandelions and, 20
    Florida beggarweed and, 70
    ragweed and, 2, 207–9
    Thoreau and, 224

"Climate-smart" agriculture, 238
Coevolution, of humans and weeds,
    generally, 2, 5–6, 14–15, 234–39
Cold War, 29, 200, 216
Columella, 109
Common ragweed *(Ambrosia
    artemisiifolia). See under* Giant
    ragweed *(Ambrosia trifida)*
Coronavirus (COVID-19) pandemic,
    235–36, 238
Cotton economy, 59–60, 63
CRISPR technology and "gene drive,"
    pigweed and, 181–82, 183
Cruydeboeck (Dodoens), 77
"Cultivate," contradictory concepts of,
    3, 127
Cyperaceae family, 108. *See also* Yellow
    nutsedge *(Cyperus esculentus)*
Cyperus papyrus, 108

Daily Inter Ocean, 61, 88
Dandelion *(Taraxacum officinale)*, 16,
    17–43
    as apomictic, 18–21, 25–26, 30, 33, 42
    common names for, 22–23
    early human cultures and, 21–23
    ecological success of, 40–41
    Freud's dandelion memory and, 31–33
    as health, 38–39
    herbicides and, 28–31, 33–38, 92
    human antipathy toward, 11, 17–18,
        24–28, 30–33
    medicinal uses of, 22
    path to weediness, 2, 40–41
    spread as humans moved westward,
        23–24
    TK (Taraxacum kok-saghyz) and,
        41–43
Dandy Blend Instant Herbal Beverage
    with Dandelion, 38–39
Darlington, William, 13, 14, 81, 82, 83,
    193
Darwin, Charles, 10, 28
David (farmer), 96–97
Davis, Fanny-Fern, 29

DDT, 36
Desmodium genus, 46. *See also* Florida beggarweed *(Desmodium tortuosum)*
DeWet, J.M.J., 10
Dewey, L. H., 118
De Witt weed, 81
Dicamba herbicide
human health and, 37
vacating of registration, 180–81
volatility of and damage to neighboring crops, 172–80
Dickinson, Emily, 24, 25
Dictionary (Miller), 77
Dinitroanaline herbicides, 118
Dinoseb herbicide, 50, 65–68, 92
"Dixie ticktrefoil." *See* Florida beggarweed *(Desmodium tortuosum)*
Dodoens, Rembert, 77
Douglass, J. L., 87
Dust Bowl, and farmers' responses to government programs, 133–34, 137
Dwayne (farmer)
foxtail and, 222–23, 224
ragweed and, 191, 202–5

Earthworm *(Lumbricus terrestris),* ragweed and, 202
Edson, Marcellus, 63
Emerson, Ralph Waldo, 4, 196
Environmental Protection Agency (EPA), 66
EPSPS (enzyme blocked by glyphosate), 142–44, 149, 171
Esculentus, 108–9
Evolved increased competitive ability (EICA), of ragweed, 206–7

Faber, Ernst, 215, 216
Faulkner, Edward, 133, 137
Ferguson, William C., 215–16
Fernald, Merritt Lyndon, 118, 216
Fertility problems (human), herbicides and, 38

"First Dandelion, The" (Whitman), 33
First Ohio Weed Book, The (Selby), 13, 14
Florida beggarweed *(Desmodium tortuosum), 44,* 45–70
as forage crop, 58–62, 69–70
herbicides and, 50, 64–69, 92
medicinal uses of, 59
other names for, 70
path to weediness, 2
peanuts and, 47–53, 63–69
peanuts and triangle trade, 53–57
Foxtail. *See* Giant foxtail *(Setaria faberi)*
France, 198, 199, 208
Franklin, Benjamin, 160, 193, 229
Freedom to Farm, 146, 148–49, 151
Freud, Sigmund, 31–33

Gene amplification, glyphosate resistance and, 171–73
Géographie botanique raisonnée (Candolle), 56
Georgia, Florida beggarweed and peanuts, 47–54
Georgia Weekly Telegraph, 61
Gerhard, Frederick, 90
Germany, 23, 198
Ghana, 45–46, 70, 122–23
Giant foxtail *(Setaria faberi), 210,* 211–33
agrestal selection, 227, 231, 232
agricultural practices and, 228–30
Amish and English farm management and, 14, 225–27
as both grains and pests, 211–12
Lenape and, 230–31
origins and spread of, 214–17
path to weediness, 2–3, 216–17
phenology of germination, 224–25
plasticity and, 213–14, 220–21, 232, 231
predictive models of germination needed for farmlands, 212–13
seedbank study and prediction of germination times, 217–25, 228–30

Giant ragweed *(Ambrosia trifida)*, 184,
   185–209
   agrestal selection, 199, 204
   allergenicity, hay fever, and pollen of,
      186, 187, 189–91, 194–98, 207–8
   climate change and, 2, 207–9
   common ragweed compared, 185–87,
      188, 192, 193, 199, 201, 208
   described, 186
   ecosystem disturbances and
      establishment of, 187–89
   evolved increased competitive ability
      (EICA) of, 206–7
   global dispersal of, 198–201, 205–7
   as "hyperaccumulator" of heavy
      metals, 205–6
   North American agriculture and,
      191–94, 202–5
   other names for, 187
   path to weediness, 2
   as "pioneering," "colonizing," and
      "keystone" plant, 187–88
   US grain exports and, 205–6
Glyphosate herbicide
   as blocker of EPSPS enzyme, 142–44,
      149, 171
   marestail and, 135, 139–40, 141
   as nonselective herbicide, 142–43
   no-till farming and, 135–36
   pigweed and, 165–66, 169–74, 177,
      180–81
   ragweed and, 204
   RoundupReady soybeans and, 145–46
   velvetleaf and, 95–97
GMO technology
   Africa and, 117
   anti-GMO movement, 154–55
   food policies and, 237
   glyphosate and, 95–96, 142–55,
      165–66, 169–72, 175–80
   non-GMO products, 12, 39
Goldenrod *(Solidago* spp), 187
Good Housekeeping, 60
Gramox. *See* Paraquat herbicide
Gray, Asa, 118

Gray's Manual (Gray, Robinson, and
   Fernald), 118
Green foxtail *(Setaria viridis)*, 211,
   214–15, 217, 230
Gremillion, Dr. Kristen, 192

Hamilton, Alexander, 74
Harlan, Jack R., 10
Harper's Weekly, 60
Hawthorne, Nathaniel, 196
Hayfever. *See* Pollen, giant ragweed and
   allergenicity
Hecate, 22
Hedysarum, 55, 58, 59
Hemp *(Cannabis sativa)*. *See* Velvetleaf
   (Abutilon theophrasti)
"Hemp brake," 80
Herbicides, generally
   biological control as alternative to,
      124–25
   no-till farming and, 133–37, 139–40,
      143
   safety issues of, 33–38
   selectivity and, 64–65, 69
   see also specific chemicals
Herms, Dan, 224–25
Herrmann, Wolfgang, 215–16
Hooker, William Jackson, 55–56
Howson, Henry, 85–86
Hungary, 98, 200

Indian mallow, 78, 81
Innovative Farmers of Iowa, 136, 153
Integrated Pest Management (IPM),
   66–69, 125

Jackson, Andrew, 58
Japan, 198, 199
Jefferson, Thomas, 24–25, 57, 78, 79
Johnson, Edward, 193
Jones, Alan Curtis, 178–79
Josselyn, John, 77

Kalm, Peter, 23, 77, 160
Kellogg, Dr. John Harvey, 63

Knot-root foxtails *(Setaria parviflora)*, 211, 214–15, 230
Knox, Henry, 97
Korea, 199

Ladies Home Journal, 35
Latham, John, 58–59
Lawns
  evolution of dislike for dandelions and, 24–25
  health and safety hazards of herbicides, 33–38
Lee, Robert E., 60
LeFranc, Emile, 87
Lenape (Delaware) people, 230–31
Linneaus, Carl, 76, 78, 160, 187, 214
Long, Bayard, 216
Longfellow, Henry Wadsworth, 25
Lunar cycles, alternative farmers and, 137, 139

Mackenzie, Morell, 195
Madison, James, 79
Mallow family (Malvaceae), 75–78, 81, 85, 90. *See also* Velvetleaf *(Abutilon theophrasti)*
Marestail *(Conyza canadensis)*, 130, 131–55
  agrestal selection, 140, 147–48, 151
  alternative farmers and, 136–39, 152–54
  anti-GMO movement and, 154–55
  biotechnology and genetics of plants, 140–42
  GMO crops, epigenetic change, and development of resistance to glyphosate, 142–55
  herbicides and no-till farming, 133–37
  other names for, 132–33
  path to weediness, 2, 132–33
Marsh, Dr. Elias J., 195
McConnell, James, 85, 86
McCormick, Robert, 80
MCPA herbicide, 37

McRae, John W., 61
Miller, Philip, 77
Modern pigweeds, 159–61, 164, 171, 180–82. *See also* Palmer amaranth *(Amaranthus palmeri)*; Waterhemp *(Amaranthus tuberculatus)*
Monroe, James, 79
Monsanto Company, 93, 97, 142–43, 145–47, 151. *See also* Glyphosate herbicide
Muhlenberg, Henry, 78
Mutagenic effects, of herbicides, 37

National Institute for Occupational Safety and Health, 38
Natural world
  evolution of weeds in response to human's alteration of environment, 234–39
  weeds and resilience of, 231–33, 238
Nature's Way dandelion root capsules, 39
Niger flora (Hooker), 55–56
Night tillage, 166–69, 181, 183
Nixon, Nicholas, 60
N. L. Willet Drug Company, 62
North Carolina State Fair, 61–62
No-till farming, 133–37, 139–40, 143, 146, 148–49
Nutsedge. *See* Yellow nutsedge *(Cyperus esculentus)*

Oliver, Mary, 232
Organic Dandelion Leaf Powder, 38
Outcrossing sexuality
  dandelions and, 19–20
  pigweed and, 160–61, 171, 181–82
  ragweed and, 204

Palmer amaranth *(Amaranthus palmeri)*. *See* Pigweed
Panicum genus, 214
Pappus, of dandelion, 19, 26
Papyrus, 108
Paraquat herbicide, 111–13, 122, 123, 125, 149

Peanuts *(Arachis hypogea)*.
 See Florida beggarweed
 (Desmodium tortuosum)
Phenology, and prediction of seed
 germination, 224, 228–30
Phytochromes, pigweed seed germination
 and, 166–69
Pickering, Charles, 55–56
Pigweed, *156*, 157–83
 agrestal selection, 161, 163–65, 166,
 174, 182–83
 agriculture and proliferation of in
 Mississippi Delta, 157–59
 ALS and evolving of herbicide
 resistance, 163–65
 Amish farms and, 14
 CRISPR technology and "gene drive,"
 181–82
 cross pollination and, 160–61
 dicamba volatility and damage to other
 crops, 173–80
 evolution of herbicide resistance,
 161–63, 165–66, 169–73
 night tilling germination-prevention
 tests, 166–69
 types of, 159–61
Pistillodium, of ragweeds, 190
Plasticity, and plant survival, 3, 11, 15,
 238–39
 dandelions and, 26, 34
 giant foxtails and, 213–14, 220–21,
 231–32
 ragweed and, 206
 velvetleaf and, 82, 83–84, 90–91, 95
Pliny the Elder, 22, 109
Plowman's Folly (Faulkner), 133, 137
Polemochorus plants, 198–99
Pollen, giant ragweed and allergenicity,
 186, 187, 189–91, 207–8
Purple nutsedge *(C. rotundus)*, 107–8

Ragweed. *See* Giant ragweed *(Ambrosia
 trifida)*
Randel, William Peirce, 223

Reciprocity, Lenape people and land,
 230–31
Republic of Tea, 39
Robinson, Benjamin Lincoln, 118
Romans, Bernard, 118
Rominger, James, 217
Roosevelt, Franklin, 91, 133
Rope, national sovereignty and need for,
 73–81
RoundUp. *See* Glyphosate herbicide
Rubisco enzyme, photosynthesis and,
 207, 214
Russian hemp, 74, 77, 80

Science, 197
Science Digest, 35
Scotts Company, 30
Selby, A. D., 13, 14
Selection pressures. *See* Agrestal selection
Setaria, 11
 Setaria adhaerans, 214
 Setaria geniculata, 214
 see also Giant foxtail *(Setaria faberi)*;
 Yellow foxtail *(Setaria pumila)*
Sida abuliton, 76, 78, 81. *See also*
 Velvetleaf *(Abutilon theophrasti)*
Slavery
 peanuts, beggarweed, and triangle
 trade, 53–57
 settled agriculture and weed removal,
 8–9
Sloane, Hanes, 54–55
Soares de Souza, Gabriel, 53
Soil balancing, to control weeds,
 136–39
Southern Agriculturist, 58–59
Southern Cultivator, 59, 60, 61
Soybean crops
 Florida beggarweed and, 47, 63–64,
 70
 giant ragweed and, 201, 204–6
 GMO and glyphosate-tolerant seeds,
 143–49
 marestail and, 143–49, 151, 153–54

pigweed and, 165–66, 170, 172, 174–79
velvetleaf and, 77, 89–94, 96
Spence, Rettia Anne, 60
Stalin, Joseph, 199–200
Stinner, Ben, 136, 138–39
Stress, epigenetic changes in plants
   and, 149–50. *See also* Agrestal
   selection
Swartz, Olof, 56

Tanzania
   Florida beggarweed and, 69–70
   nutsedge and, 103–7, 110–17,
   121–28
Taraxacum species, 11, 19–20.
   See also Dandelion *(Taraxacum
   officinale)*
Tariffs, on rope materials, 74–75,
   79–80
Theophrastus, 22, 76, 109
Thermo-hydro-oxy-driven process, of
   seed germination, 222
Thoreau, Henry David, 196, 223–24,
   232, 233
Tilling (hoe culture), and its
   consequences, 7, 116
   dandelions and, 27
   Florida beggarweed and, 51, 63
   foxtail and, 220–21
   giant ragweed and, 202
   lunar cycle and, 137, 154
   marestails and, 132–34
   no-till farming, 133–37, 139–40, 143,
   146, 148–49
   nutsedge and, 110, 113
   pigweed and night tillage, 166–69,
   181, 183
   selection pressures and, 10
   velvetleaf and, 83, 94
   yellow nutsedge and, 104, 118
TK (Taraxacum kok-saghyz), dandelions
   and, 41–43
Tompkins, Daniel F., 87
Triazines, pigweed and, 162–63

Trichomes, Florida beggarweed and, 52
2,4-dichlorophenoxyacetic acid (2,4-D),
   92, 118
   dandelions and, 28–30, 34, 36–37
   foxtail and, 216–17
   health and safety issues of, 36–38
   pigweed and, 162
   ragweed and, 197

United Nations Millennium Development
   Goals, subsistence agriculture and,
   116–17
USDA
   Division of Agrostology, 62
   Florida beggarweed and, 70
USSR, ragweed and, 199–200

Van Buren, Martin, 58
Vavílov, Nikolai, 75
Velvetleaf *(Abutilon theophrasti)*, 72,
   73–100
   agrestal selection and, 84, 92,
   98–100
   ancient uses for, 75–76
   attempts to industrialize production
   of, 85–89
   common names for, 76, 77, 81
   hemp and spread of, 74–80
   herbicides and, 91–98
   medicinal uses of, 89
   path to weediness, 2
   plasticity of, 83–84, 90–91, 95
   seed dormancy of, 82–83
   soybean crops and, 89–94
   U.S. sovereignty and need for ropes,
   73–81, 85–89
Vogel, Theodore, 56
*Voyage to the Islands Madera, Barbados,
   Nieves, S. Christophers, and Jamaica*
   (Sloane), 54–55

Walden (Thoreau), 223–24
Wallace, Mike, 178, 179
Washington, George, 75–77

Waterhemp *(Amaranthus tuberculatus)*,
    14, 160–61, 164–65, 169–71, 182
Waterhouse, Sylvester, 87
Webster, Daniel, 196
Webster, Ted, 96, 97
"Weediness," paths to, 1–15
    agriculture, slavery, and weed removal,
        8–9
    agriculture, weeds, and human
        settlements, 6–7, 11–13
    agriculture and agrestal selection,
        9–11, 13
    plant evolutionary biology and human
        behavior, 1–6, 13–15
    "weedy traits" of humans, 239
    "weedy traits" of plants, 4–5, 84, 98
Weed biology
    allelopathy, 119–20, 168
    breeding system, 19–20, 46, 161,
        181–82, 190, 213
    coevolution with humans, 2, 5–6,
        14–15, 234–39
    competition, 14, 64, 83–84, 94
    dormancy, seed, 82, 83, 219–22
    dormancy, tuber bud, 113–14
    genetics, 11, 20–21, 143, 147–50,
        165, 171
    plants of contradiction, 3
    plasticity, 26, 83, 90, 213–14, 220,
        232, 236
    population growth, 81–84, 93
    seed dispersal, 9–11, 26, 52, 55, 70,
        108, 166, 187, 198, 200, 239
    seed germination, 82–83, 166–68,
        219–23, 228

seedbank, 88, 168, 191, 217–23, 226,
    231
succession, 134
Whitman, Walt, 33
Wilson, Woodrow, 198–99
Women, labor of weeding and dangers of
    herbicides, 104–5
Wood, Fernando, 60
World War I, ragweed and, 198–99
Wyman, Dr. Morrill, 194, 195

Yellow foxtail *(Setaria pumila)*, 210,
    211, 227, 230. *See also* Giant foxtail
    (Setaria faberi)
Yellow nutsedge *(Cyperus esculentus)*,
    102, 103–28
    agrestal selection, 109, 114
    agriculture in North America and,
        117–19
    allelopathy and, 119–20
    biological control as alternative to
        herbicides, 124–25
    described, 108
    herbicides and, 104–5, 110–13, 118–19,
        121–22
    path to weediness, 2
    pesticides as "medicine" and, 115–16
    reproduction of, 108–9, 113–14
    Tanzanian farmers and "nati nyasi,"
        103–7, 110–17, 121–28
    types and morphs of, 107–10
    see also Chufa ("tiger nut")
Yugoslavia, 200

Zeisberger, David, 230